Studies in Applied Philosophy, Epistemology and Rational Ethics

Volume 55

Studies in Applied Philosophy, Epistemology and Rational Ethics (SAPERE) publishes new developments and advances in all the fields of philosophy, epistemology, and ethics, bringing them together with a cluster of scientific disciplines and technological outcomes: ranging from computer science to life sciences, from economics, law, and education to engineering, logic, and mathematics, from medicine to physics, human sciences, and politics. The series aims at covering all the challenging philosophical and ethical themes of contemporary society, making them appropriately applicable to contemporary theoretical and practical problems, impasses, controversies, and conflicts. Our scientific and technological era has offered "new" topics to all areas of philosophy and ethics—for instance concerning scientific rationality, creativity, human and artificial intelligence, social and folk epistemology, ordinary reasoning, cognitive niches and cultural evolution, ecological crisis, ecologically situated rationality, consciousness, freedom and responsibility, human identity and uniqueness, cooperation, altruism, intersubjectivity and empathy, spirituality, violence. The impact of such topics has been mainly undermined by contemporary cultural settings, whereas they should increase the demand of interdisciplinary applied knowledge and fresh and original understanding. In turn, traditional philosophical and ethical themes have been profoundly affected and transformed as well: they should be further examined as embedded and applied within their scientific and technological environments so to update their received and often old-fashioned disciplinary treatment and appeal. Applying philosophy individuates therefore a new research commitment for the 21st century, focused on the main problems of recent methodological, logical, epistemological, and cognitive aspects of modeling activities employed both in intellectual and scientific discovery, and in technological innovation, including the computational tools intertwined with such practices, to understand them in a wide and integrated perspective.

Studies in Applied Philosophy, Epistemology and Rational Ethics means to demonstrate the contemporary practical relevance of this novel philosophical approach and thus to provide a home for monographs, lecture notes, selected contributions from specialized conferences and workshops as well as selected Ph.D. theses. The series welcomes contributions from philosophers as well as from scientists, engineers, and intellectuals interested in showing how applying philosophy can increase knowledge about our current world. Initial proposals can be sent to the Editor-in-Chief, Prof. Lorenzo Magnani, lmagnani@unipv.it:

- A short synopsis of the work or the introduction chapter
- The proposed Table of Contents
- The CV of the lead author(s).

For more information, please contact the Editor-in-Chief at lmagnani@unipv.it.

Indexed by SCOPUS, ISI and Springerlink. The books of the series are submitted for indexing to Web of Science.

More information about this series at http://www.springer.com/series/10087

Jean Lassègue

Cassirer's Transformation: From a Transcendental to a Semiotic Philosophy of Forms

 Springer

Jean Lassègue
Centre Georg Simmel, Recherches Franco-Allemandes
en Sciences Sociales (CNRS-UMR 8131)
École des Hautes Études en Sciences Sociales
Paris, France

ISSN 2192-6255　　　　　　　　ISSN 2192-6263　(electronic)
Studies in Applied Philosophy, Epistemology and Rational Ethics
ISBN 978-3-030-42907-2　　　　ISBN 978-3-030-42905-8　(eBook)
https://doi.org/10.1007/978-3-030-42905-8

This Springer imprint is published by the registered company Springer Nature Switzerland AG
The registered company address is: Gewerbestrasse 11, 6330 Cham, Switzerland

Preface

All the quotes from Cassirer are given in English and followed by the reference to the now standard edition, in German, published by the publishing company Felix Meiner in two series:

(i) *Gesammelte Werke*, Hamburger Ausgabe, Felix Meiner Verlag, 1998–2009, appearing in footnotes as "ECW" followed by the page number.

(ii) *Ernst Cassirer Nachgelassene Manuskripte und Texte*, Felix Meiner Verlag, 1995–2017, appearing in footnotes as "ECN" followed by the page number.

Some of Cassirer's texts were written in English (essentially those written during the last period of his life: *An Essay on Man* and *The Myth of the State* as well as various articles) and are directly quoted from the standard Felix Meiner edition mentioned above. For quotes from the other texts, written in German, I used the existing English translations when available (in which case the reference of the translation is provided along with the reference to the German edition) or I translated them myself from the German original when no translation was available.

Paris, France   Jean Lassègue

General Introduction

It is paramount, when beginning to read a new book, to contemplate its global purpose. Ours consists in an attempt to clarify the status Ernst Cassirer ascribed to the *transcendental* point of view, understood in the sense of the conditions of possibility enabling to account for the diversity of types of knowledge. Cassirer's reflections are thus intrinsically gnoseological, and epistemological considerations form a significant aspect of them. With his vocabulary and methods bearing the mark of the epistemological concerns engaged by Kant, a legacy having extended into the twentieth century through the neo-Kantian school of Marburg towards which Cassirer recognizes his debt, it is from such a perspective that he believed the question of the conditions of possibility of knowledge should be approached.

Nevertheless, the manner in which Cassirer conceives of these conditions of possibility diverges from that of his Kantian or neo-Kantian predecessors and particularly calls into question the privilege conferred to epistemology and, therein, to the analysis of the exact and natural sciences (respectively mathematics and logic on the one hand and physics and chemistry on the other). Cassirer does not define the transcendental by remaining attached to a notion of objectivity specific to these sciences nor does he conceive of their objectivity as the most momentous manifestation of rationality, thus making these sciences the models to be emulated by all other forms of knowledge, as the Kantian and neo-Kantian traditions had tended to do.

Guided by the deep changes impacting the exact and natural sciences since Kant, Cassirer was on the contrary confronted with the necessity of accounting for the possibility of their *internal evolution* rather than only accounting for the possibility of their *originating principle*. Now, for Cassirer, this internal evolution led the exact and natural sciences in their modern developments to *pluralize their own modes of objectivity*: Cassirer thus situates the issue of the variability of the modes of objectivity at the very core of the post-Newtonian sciences and his philosophical gesture consists in drawing all the consequences from the standpoint of the diversification of knowledge *in general*. This change in perspective consists therefore in going beyond the strictly epistemological point of view, *for reasons internal to epistemology*, that is, approaching epistemology not as a content restricted to the scientific domain but as the continuous transformation of the modes of objectification, a transformation which opens up to a semiotic dimension. From that point of view, Cassirer remained an epistemologist through and through. But it is according to this new epistemological perspective that the notion of the transcendental is transformed into an inquiry regarding sense, meaning and symbol which lead Cassirer to question the status conferred to the exact and natural sciences as a model for knowledge: by dissociating the transcendental as a process from the mode of objectivity specific to the scientific domain, the analysis of conditions of possibility now concerns all forms of objectification, be it the analysis of myth, law, art, technology, or any other form of knowledge or practice. This is because such a generalization of the transcendental point of view is not limited to the various fields of knowledge but extends to signifying practices in general, as is the case for instance with rituals or technology.

Valued in their own right, the forms of objectification are then no longer tied up to the exact and natural sciences conceived as the final goal of any project for knowledge: each form—in the sense of a signifying practice—deploys an intrinsic coherence according to its own specific modalities and a specific type of objectification. The fields of knowledge no longer being hierarchically organized according to their greater or lesser proximity to the exact and natural sciences in terms of object or method, it is henceforth the question of the variability of their mode of objectification and of their reciprocal relationship which becomes the focus of inquiry.

It is therefore crucial to avoid a major pitfall into which it would be easy to be trapped. It would consist in considering that there is no longer any objectivity whatsoever, since there is no longer any guaranteed universal grounding in the mode of objectivity specific to the objective sciences; the very idea of science would then become questionable. Relativism thus looms behind any attitude which, by seeking to attribute relevance to the variability of modes of objectification, would finally tend to equalize this variability due to the absence of a unique mode of objectivity, thereby producing the inversed image of objectivity, but only in a sceptical fashion. Cassirer denounces this sceptical attitude which renounces the exploration of the possibility that the transcendental may itself be the motor for the variability of its own modes of objectification.

Leaning towards dogmaticism, both objectivistic and relativistic attitudes must therefore be overcome in order to enable a *variational* interpretation of the forms of objectification and to account for what enables each of them to not only ensure the particular mode of constitution and institution of their specific object, but also to develop these specific modes of constitution and institution *over the course of time*. The conditions of possibility specific to the exact and natural sciences thereby become resources, indispensable but partial, for thinking about the more general problem of the constitution of knowledge and practices in all of their forms. Cassirer dismissed the traditional interpretational framework of epistemology and thereafter attempted to renew it by making use of two notions: on the one hand, the distinction between *substance and function* and on the other hand, the notion of *symbolic form*.

Distinguishing between substance and function enabled Cassirer to characterize the exact and natural sciences as well as their evolution. Science starts by considering the objects it determines as ready-made and existing in an external nature, i.e. as substances. But the inner evolution of science would gradually steer away from this naïve realism and conceive its own objectivity in a purely functional manner as a system of ideal elements of which the relationships are subject to interpretations that could even consist in putting them into algebraic form. The substance/function distinction thus plays a double role which is both historical and gnoseological: from a historical standpoint, it enables to distinguish between antiquity and medieval periods on the one hand and modern and contemporary periods on the other, enabling to direct the attention towards specific moments during which a shift from one conception to the other occurred, such as the Renaissance. From a gnoseological standpoint, it enables to demonstrate that naïve realism of the substantial type is never completely eliminated by the ulterior development of a form of objectification and that the functional point of view is never fully entrenched but consists in a continuous process of distancing itself from a substantial conception which always precedes it.

Just as the distinction between substance and function enables to study the dynamics of science from an historical point of view, so does the notion of a "symbolic form" with the new epistemological framework Cassirer elaborates. The notion of a "symbolic form", for its part, constituting Cassirer's major contribution to philosophical thought, involves, in order to be fully grasped, a detachment from the classical oppositions of epistemology—those of the necessary and of the contingent, of the *a priori* and of the empirical, or of the rational and of the irrational—these being notions having all been elaborated in an interpretational framework within which objective science still served as paradigmatic value for rational knowledge. The notion of a "symbolic form" does not lie within these strict divisions, which are too rigidly set to begin with, and it rather enables to *dynamically* study the institution and evolution of forms of objectification, their relatedness, their divergences and persistence which are not limited to a circumscribed type of knowledge proper. The notion of a "symbolic form" therefore makes room for the notion of *culture* by making the specific dynamic of culture rest on the variability and the intrinsic deployments of forms conceived as schemas that generate the aforementioned modes of objectification: language, myth, art, technology, law, etc., that are found simultaneously in forms of knowledge and activities.

Epistemology as it is renewed by Cassirer is certainly faithful to the Kantian research project as far as it rests upon the analysis of conditions of possibility of knowledge, but it is also deeply original with respect to this programme in so far as it significantly reworks the modalities of its realization. This because the two new notions that Cassirer elaborates (the difference between substance and function and the notion of a "symbolic form") no longer place interiority—be it that of the transcendental subject—but rather *symbols*—that is, immediately public semiotic materials—at the very core of the elaboration and of the transformation of the forms of objectification. This represents a major change in the practice of transcendental inquiry as it reconstitutes the whole project from within because if the conditions of possibility of knowledge are no longer ascribed to the inner realm of a transcendental subject, they proceed from social conditions designated as *cultural processes*.

Grounding cultural processes in the trans-individual and directly social dimension of symbolism certainly presents the danger of ignoring that interactions are also made by individuals. But it is not only at an individual level that culture presents itself in concrete situations. Any individual will first *inherit* cultural schemas embodied in symbols that will be repeated as closely as possible—such is the case with respect to one's native language, for instance—before the individual is able to demonstrate some form of originality and to transmit one's own modified version of the cultural schema which was in use. The intertwining of individual and trans-individual levels is therefore continuously at work, and this has deep epistemological consequences.

Effective symbolic practice as it is inherited and transmitted is thereby neither necessary nor contingent and does not constitute a stabilized knowledge that can be accumulated and transmitted. It pertains neither to a "pre-conceptual" nor to an "ante-predicative" sphere which would only reaffirm a strict distinction between knowledge and non-knowledge. The notion of a "symbolic form" on the contrary enables to adopt another point of view regarding the processes of construction and institution of the modes of objectification. From this point of view, the notion of symbol as it is used by Cassirer may at first baffle the reader owing to its extreme plasticity: as Cassirer himself remarks, the term "symbolic"—which he most of the time employs as an adjective—can have a magical, allegorical or even algebraic meaning. But this is precisely the reason why the adjective "symbolic" is of interest for an undertaking that aims to understand *how the production of meaning encompasses its own variability* as is actually shown by the variety of forms of objectification.

Leaving aside the focus on the monosemic foundation of knowledge as the most fundamental task of philosophy, the syntagm "symbolic form" has the ambition of objectively accounting for the progressive diversification of the modalities of meaning and of the mutual relationships between these modalities. This is a major contribution of Cassirer's approach which profoundly modifies the divisions between the disciplines, in particular with respect to the difference between the exact sciences, the natural sciences, and the humanities and social sciences, because symbolic forms must be considered as thematic motifs which traverse the multiple strata of knowledge and practices while ignoring disciplinary enclosures. The modalities of these variations thus trace the contours of what can be called *cultures* whose forms can therefore be analysed in terms of dissemination, stability and persistence.

Semiotic approaches become therefore the core of the transcendental analysis of the conditions of possibility of knowledge understood as the conditions of possibility of meaning: what constitutes the core of the inquiry thereafter has for object the *permanent transformation* of meaning through the diversification of the forms of objectification, this being a form of generativity specific to meaning which constantly exceeds itself and of which the notion of a "symbolic form" seeks to understand the dynamic.

Keeping in mind the new dynamic turn of the transcendental enquiry, the problem of the goal of the whole process of meaning transformation is therefore left open: towards what end does it tend? The answer to this question is delicate and controversial because it touches upon the very nature of Cassirer's approach. It should first be noted that his project regarding a "philosophy of symbolic forms" varied during the time it was being carried out, this providing a nice example of the renewal and transformation of meaning when inscribed within a practice: starting from a quasi-Hegelian point of view which established a strict correspondence between the internal development of a "symbolic form" and an organized distribution of "symbolic forms" in a sequence ranging from the most expressive to the most objective, Cassirer later implicitly renounced such a schema whereby the "symbolic forms" were strictly ordered along an axis oriented towards the advent of the objective concept. The question of the closure towards which the process generating the forms of objectification tends is then posed with heightened perspicacity.

Emphasizing the internal transformation of Cassirer's project made it methodologically necessary for our argument to introduce a somewhat genetic perspective which would enable to account for its evolution without seeing contradictions. If the process generating symbolic forms is no longer considered to be oriented towards a conceptual end, the question that needs to be answered is whether it is necessary to suppose the existence of a governing function attached to such a generative process. The existence of such a function seems to be indispensable for avoiding an integrally empirical approach which would involve renouncing the anticipatory aspect present in the notion of transcendentality. Cassirer scholars would generally answer the question by attributing such a governing function to language or even to a more general "symbolic function"—an expression that Cassirer seldom makes use of—and by seeing in it the condition of possibility of the advent of the Subject, conceived as the ultimate end of the transcendental process. We have not adopted such a perspective here because, in addition to its difficulty in accounting for Cassirer's corpus as a whole, it tends to make the process which generates forms of objectification subordinate to individual psychology, although this process should be thought of as an essentially cultural one which is social right from the onset.

Rationality should still be defined as a structure of anticipation but how could it be understood as accounting for the production of forms according to an order without projecting a final end, be it conceptual or subjective? Posed in such a manner, this question, clearly in line with the Kantian and neo-Kantian reflections regarding the nature of transcendentality, amounts to knowing whether it is possible to uncover *categories* that anticipate the possible modalities of objectification. The difficulty of responding to this question stems from the fact that the aim of the research carried out by Cassirer does not focus on the univocal determination of objectivity in a system of fixed categories, but rather on the dynamic process of the forms of objectification.

Despite the fact that Cassirer never proposes a closed list of categories, he never renounces a transcendental approach to rationality as a philosophical method.

We can therefore suppose that the problem of the existence of categories shifted due to the dynamic nature attributed to the process generating forms of objectification. And indeed, since the forms of objectification are never entrenched with respect to a particular type of signifying practice but are disseminated across all fields of culture, the process generating these forms must be found in *transcategorical operations* the dynamic nature of which needs to be described. These transcategorical operations are at the centre of what induces permanence and variation within cultural processes and will be analysed in detail.

Effects of these transcategorical operations generate cultural fields that Cassirer studied in their unlimited variety: it is impossible, from this point of view, not to be struck by the extraordinary profusion of domains that Cassirer addressed in his works, from relativistic and quantum physics to Renaissance philology, from comparative grammar and mythology to nascent structuralism, and from the study of Renaissance philosophy to the study of Cambridge Platonism. And yet, some results or methods which have now been integrated into the edifice of knowledge, such as Weber's sociology or Freud's psychoanalysis, have barely been addressed by Cassirer[1] or, at least, have not been specifically thematized as such. In this lies something of a predicament which could impair the recognition of the profound topicality of his work.

Grounded in an immense erudition carried out to the forefront of the research of his time, Cassirer's works have nevertheless suffered from the particular circumstances related to his personal exile as well as the dispersion of German-speaking intellectuals during the Nazi era that contributed to their delayed recognition. But today, these historical circumstances should be viewed more as an incentive than as an obstacle to the pursuit of the collective constitution of knowledge in the broader sense that Cassirer so forcefully advocated and put into practice. This is what we shall advocate in further detail now.

---

[1]Cassirer rarely cites the work of Freud, the later developments of which he was probably unaware due to the exile with which both thinkers were stricken with the rise to power of the Nazis in Germany and Austria. It is, in any case, for its anthropological more than its psychoanalytical aspects that Cassirer refers to Freud, in particular concerning the anthropological object which is so characteristic of the end of the nineteenth century and of the beginning of the twentieth century: totemism (cf. E. Cassirer, *The Myth of the State*, ECW 25, pp. 32–36 where it is question of *Totem and Taboo*) or concerning Freud's methodology in anthropology (*An Essay on Man*, ECW 23, p. 83). With respect to Max Weber, rarely does Cassirer address his works (cf. for example, *The Philosophy of Symbolic Forms*, volume 2, p. 193, footnote 41 (ECW 12, p. 227 Footnote 85). But that is most likely due to the particularity of the reception of Weber's work which took over ten years to impose itself as a major reference after the death of its author in 1920.

Contents

Part I
Epistemology

Cassirer completed his graduate studies (in philosophy but also in law and German literature) in Berlin and in Marburg where in 1899 he defended his doctoral thesis on the Cartesian theory of mathematics and the natural sciences, under the direction of Paul Natorp. The choice of the University of Marburg (in the region of Hesse) was suitable for someone who wished to pursue epistemological research since a current of thought, which has since been called "The Marburg School", had constituted itself around two professors, Hermann Cohen (1842–1918) and Paul Natorp (1854–1924): this current aimed in particular to rekindle with the epistemological aspects of Kantianism, beyond the great systems of German idealism—in particular that of Hegel—by taking the full measure of the scientific developments having taken place during the nineteenth century in the fields of mathematics and of the natural sciences.

It is within this tradition that the epistemological work of Cassirer falls: basing itself on what his master's at Marburg called the "fact" of science—a universal and necessary fact—it was a matter of understanding its nevertheless *historical* nature, not only as regards the slow accumulation of objective contents but also with respect to the translation of philosophical categories into regional concepts such as operated within the sciences, for instance in mechanics, where the category of substance is translated into the "regional" concept of mass or where the category of causality is translated using the concept of force. From this standpoint, Cassirer's epistemological elaboration seeks to further the legacy of the Marburg school, even if he is confronted with a whole different epistemological situation, dominated by the fact that, henceforth, the translation of a philosophical category would operate according to several possible "regional" concepts, as in the case of the diversity of geometries. This problem of the variety of regional concepts within science itself would have, for Cassirer, considerable philosophical consequences because it led to conceive of the *possibility of the transformation of philosophical categories themselves and of their reciprocal relationships,* as for example in the case of the relationship between "substance" and "function". Also, the problem to which Cassirer was confronted was broader than that put forth by the Marburg school and Cassirer recognized his debt towards other philosophical currents, be they those that were rich in epistemological content such as Leibnizian philosophy during the classical period or those

of a more metaphysical orientation stemming from German Idealism and its historical reflections, particularly with Hegel. This double heritage, both epistemological
and philosophical, also enabled Cassirer to guard himself against any form of *relativism*—which would have interpreted the possibility of a historical evolution of the
categories of thought as the negation of the very possibility of science—all the while
providing for the possibility of a history of science. Hence, we understand that it is
not only the renewal of scientific contents which may pose a problem but rather the
transformation of the very categories of thought inasmuch as it is at such level that
the issue of the nature of their historical temporalization intervenes.

It is within this problematic framework that Cassirer would attempt to renew Kantianism. It would first mean to get to grips with the "history of pure reason", which
Kant had outlined in the end of the *Critique of Pure Reason*, but which he thought
to be able to treat as a subsidiary task to be eventually undertaken and which did not
fundamentally put the achievements of his philosophy into question. But for Cassirer,
it was necessary to confront the "fact" of science in the historical transformation of
categories which this fact implied, in order to renew what was to be specifically
understood by "fact" of science and to thus preserve the Kantian developments concerning the central character, for modern rationality, of a mathematical theory of
nature. It was thus necessary to begin by recognizing where the perspective opened
by Kant appeared as outdated, philosophically and epistemologically—both points
being linked of course.

Philosophically, already in the very first volume of his historical epistemology
work on modern science (*The Problem of Knowledge in Modern Philosophy and
Science*), in 1906, Cassirer posed the diagnostic by declaring that it is the rigid
character of the Kantian system of categories that poses problems for whom wants
to take the current dimension of the "fact" of science into account[1]:

*"The "fact" of science is and stays of course, according to its own nature, a fact
which develops historically. While in Kant this insight has not yet been recognized
unambiguously, while categories, as far as their number and content are concerned,
can still appear in him as ready-made "fundamental concepts of the understanding",
the modern enhancement of critical and idealistic logic has brought full clarity on this
point. The forms of judgement only mean unifying and living motifs of thinking which
traverse through the whole diversity of its specific forms and mobilize themselves in
creating and expressing always new categories. The more these variations are proved
to be rich and flexible, the more they confirm the specificity and originality of the
logical function from which they emerge."*

Epistemologically, Cassirer poses his diagnosis in two stages. Firstly, by discovering, during his student years, the mathematical revolution operated by Felix Klein
which acknowledges the plurality of geometries and finds the mathematical means
to classify them using group theory, Cassirer takes heed of the changes which must

---

[1] E. Cassirer, Das Erkenntnisproblem in der Philosophie und Wissenschaft der neueren Zeit, Band
1, ECW 2, pp. 14–15. From now on, the first three non-translated volumes of Cassirer's Das
Erkenntnisproblem in der Philosophie und Wissenschaft der neueren Zeit will be quoted in an
abridged form: "E. Cassirer, Das Erkenntnisproblem...", followed by the number of the volume
("Band") and the reference to the German edition ("ECW").

now affect the Kantian notion of "pure intuition", itself lying at the basis of the very possibility of an *a priori* mathematical science. Then, by discovering, between 1910 and 1920 the depth of the incidence of this new geometrical framework upon the very destiny of physics since it makes possible Einsteinian relativity, Cassirer measures the extent to which the philosophical change entailed by the science of his era not only pertains to the mathematical usage of categories, but also to their employment in physics. Thus, on the one hand it is the internal evolution of the axiomatic undertaking in geometry and on the other hand, its consequences extending to relativistic physics which have completely modified the relationship between the theoretical categories of philosophy and the exact and natural sciences.

Let us thus begin by providing a brief overview of these two perspectives, the one being philosophical and the other being epistemological, and of their consequence upon the progressive elaboration, by Cassirer, of a reflection pertaining to the more general notion of the constitution of meaning.

Chapter 1
The Epistemological Situation of Cassirer

Abstract Contrary to what is sometimes claimed, there is no real "turn" from philosophy of science to philosophy of culture in Cassirer's intellectual development: it is the internal evolution of Geometry and its consequences in Physics that lead Cassirer to broaden an essentially unique perspective. Before Cassirer became intellectually active, the emergence of non-Euclidean geometries had triggered a crisis in Mathematics that had consequences in Natural sciences as well as in Philosophy. The concept of group of transformation in Geometry (Klein) and the subsequent use of Riemanian geometry in relativity theory (Einstein) were responses to this crisis in Mathematics and Physics, respectively. The philosophical response to the crisis prompted Cassirer's original concept of a "Symbolic Form". From a transcendental perspective, his approach would transform the concept of objectivity into that of the various modes of objectivation. Thus, Cassirer's philosophical project can be stated as follows: since some modes of objectivation that pertain to Mathematics have also meaningful consequences in Physics, it is possible to philosophically consider the very notion of the variety of modes of objectification and their possible transfers as the key concepts of meaning formation in general, thus transforming the transcendental perspective into a semiotic one. According to Cassirer, this transformation becomes the goal of the philosophical enquiry and can be studied in all cultural forms: Language, Mythical thinking, Law, Art or Technology, to name a few.

Keywords Non-Euclidean geometries · Group of transformation · Relativity theory · Felix Klein · Geometry and philosophy · Transcendental perspective

In this chapter, we shall attempt to answer the following question: what led Ernst Cassirer, who was initially an epistemologist of the exact and natural sciences, to broaden the scope of his investigations towards the more general issue of the construction of meaning, without limiting himself to its purely scientific construction?

This broadening of the scope of Cassirer's investigations is generally ascribed to his discovery in 1921 of the "Library for Cultural Studies" constituted by Aby Warburg whose unprecedented and evolutive classification order challenged the established thematic categories. But this discovery only had the determining impact it did on Cassirer's career because he was able to integrate it with the prior epistemological

© Springer Nature Switzerland AG 2020
J. Lassègue, *Cassirer's Transformation: From a Transcendental
to a Semiotic Philosophy of Forms*, Studies in Applied Philosophy, Epistemology
and Rational Ethics 55, https://doi.org/10.1007/978-3-030-42905-8_1

work which had prepared him towards that day of 1921. The answer to our opening question therefore does not depend upon a simple contingent fact, as important as it may be, but proceeds from an evolution internal to the sciences of Cassirer's time. To formulate a response to our question: the opening towards the general issue of the construction of meaning is a *consequence of the evolution internal to geometry and to its repercussions in physics*.

In order to justify this response, as we shall endeavor to do in the following pages, it is necessary to begin with a historical remark which will divide Cassirer's itinerary into two periods. It is starting in 1920–1921 that Cassirer took measure of the capital epistemological change provoked by the transformations that the modern conception of geometry would have upon his own philosophical project. Before this date, that is, between 1906 and 1920, Cassirer's line of inquiry remained, globally, the very one he had been following since his studies in Marburg at the end of the nineteenth century: from an epistemological point of view, it was a question of tackling the philosophical issue of the historicity of the "fact of science" by justifying the possibility for a historical deployment of pure reason, in particular, starting with the advent of the mathematical science of nature since Galileo and Descartes, and in keeping with what Kant had outlined in the chapter "The History of Pure Reason" at the end of the *Critique of Pure Reason*. However, starting in 1920, things began to change: Cassirer became aware that the mathematical concept of *group* not only bore capital epistemological significance with respect to the transformation of geometry and physics after Kant but also that this concept would require from him a profound rethinking of his initial philosophical project. This transformation consists, as we will later see in detail, in taking the full measure of the solution provided by Felix Klein regarding the problem of the plurality of geometries (Euclidean and non-Euclidean) which he nevertheless managed to regroup under the unity afforded by the concept of group in projective geometry: Cassirer saw in this the very example of the construction of meaning in general, be it scientific or not. It would henceforth be necessary to understand by "construction of meaning" not the conditions of objectivity or of the "fact of science", but to see in variations of meaning the *conditions of its possible invariance through time*. In order to understand how meaning endures through time, it is not a question of removing oneself from time by relying upon an immutably set system of categories; on the contrary, it is necessary to delve into the variations in order to reveal a meaning which endures, the variability of categories becoming the very sign of the progressive emergence of an invariant. Therefore, it is necessary to *always presuppose* the meaning and to interpret its internal transformations as an *indication of the invariability of its form* through time.

Accordingly, there is a veritable turning point in Cassirer's path, which he had not directly thematized himself, even if it is possible to reconstitute it with a certain probability.[1] This turning point heralds the apparition of his philosophy's key concept, that of "symbolic form", which he began to elaborate as early as 1921 and which

---

[1]Two concordant elements enable to justify this interpretation: (i) the publication in 1921 of his book about Einstein (E. Cassirer, *Einstein's Theory of Relativity*, The Open Court Publishing Company, Chicago, 1923, ECW 10) in which he demonstrates the capital role played by non-Euclidean geometries and the concept of group in the constitution of the Einsteinian theory of relativity; and

he will explore in his most famous book, *The Philosophy of Symbolic Forms*, of which the three volumes were published respectively in 1923, 1925, and 1929.[2] It is indeed by basing himself upon an epistemological reflection regarding the role of the concept of group in the solution to the problem of the multiplicity of geometries, a solution which was provided by the concept of symbolic form, that Cassirer will first find his philosophical inspiration.

Let's now address the details of the analysis of the problem encountered by Cassirer with respect to the transformations of geometry throughout the first half of the nineteenth century. In what way did the disorderly state of geometry pose a specifically philosophical problem?

1 Geometry and Philosophy

1.1 The Definition of Rationality at the Core of the Relations Between Geometry and Philosophy

In order to answer the question above, it is necessary to make a small digression and to give a few indications regarding the role played by geometry in the general relationships existing between mathematics and philosophy. Placing this question within Cassirer's perspective will require three remarks from a cultural, epistemological, and philosophical standpoint.

First of all, from a cultural standpoint, geometry has played a fundamental historical role at the very origin of Western culture in Ancient Greece: geometry forms part of a heritage enabling to conceive of the present in historical continuity with such heritage, conversely to other cultural forms such as religion which clearly removes

---

(ii) the autobiographical remarks provided in Cassirer's last manuscript (ECN 8, pp. 187–188) bearing the date of his death (April 13, 1945) in which Cassirer mentions, on the one hand, the serious problem the very notion of non-Euclidean geometry represented for the Kantian student he was because it completely exceeded the framework of the *Critique of Pure Reason* and on the other hand, the solution which Felix Klein had produced with his use of the concept of group. Cassirer does not mention the date on which he read Klein's works (whose most famous text, designated as the "Erlangen Program" is dated 1872), but he only cites the 1921 edition of Klein's *Collected Works*. Even if Cassirer had already, during his student years, understood the solution Klein provided to the problem of the plurality of geometries, there are good reasons to believe that it is only around 1920–1921, the period during which he wrote his book on Einstein, that he understood the full epistemological, and more broadly, philosophical interest which could be derived from it. It is also not impossible that the interest he expressed already as a student regarding the concept of group was in fact an *a posteriori* reconstruction given that it was only much later, in 1921, that this interest manifested itself in his works.

[2] A last, posthumous, volume, regroups texts dating from 1928 to 1940 (*The Philosophy of Symbolic Forms*, vol. 4, trad. J. Krois & D. Verene, Yale University Press, 1996), is now published as the first volume in the series *Ernst Cassirer Nachgelassene Manuskripte und Texte*, Felix Meiner Verlag, 1995–2017 (ECN 1).

us from Ancient Greek culture. In the context of this historical continuity, geometry is understood as the science pertaining to the measurement of figures whereas arithmetic theories, although present in Ancient Greece as well, never enjoyed the same development, probably due to issues related to graphical representation. In any case, geometry has, from the onset, been partly linked with philosophy inasmuch as these two disciplines form the two facets of a same *theoretical ideal* of knowledge.[3] By "theoretical", we mean that the description of the world and the search for its causes are not only to be pursued in the realm of the mythical tale,[4] but that such description must adopt a register of expression along the lines of demonstrative argumentation centered upon the concept of the identity of measurements. By "ideal", what is meant is that geometry and philosophy share the same particular requirement of conceiving of measurement-based demonstrative argumentation as the *sole norm* in what concerns the description of the world and its causes.[5] Of course, demonstrative argumentation having recourse to only itself cannot do without first principles that have first been intuitively justified. Thoroughly theoretical research in philosophy concerns precisely these first principles. But, contrarily to philosophy, geometry manages to integrally define its own first principles by providing itself, already with Euclid during the third century B.C., with an *axiomatic* basis from which it deduces its propositions by means of demonstration based on first principles considered to be indispensable. Hence, from a philosophical standpoint, geometry plays an ambivalent role: on the one part, we can interpret it as the *most perfect achievement of the ideal of knowledge* and attempt to *extend its method* towards any form of knowledge

---

[3]E. Cassirer, *Substance and Function*, The Open Court Publishing Company, Chicago, 1923, pp. 268–269 (ECW 6, pp. 289–290): "The procedure of the "transcendental philosophy" can be directly compared at this point with that of geometry. Just as the geometrician selects for investigation those relations of a definite figure, which remain unchanged by certain transformations, so here the attempts is made to discover those universal elements of form, that persist through all change in the particular material content of experience".

[4]Only the fourth and last volume of *Das Erkenntnisproblem*... has been translated into English (E. Cassirer, *The Problem of Knowledge; Philosophy, Science, and History since Hegel*, transl. by William H. Woglom & Charles W. Hendel, New Haven, Yale University Press, 1950). From now on, this translation will be quoted "E. Cassirer, *The Problem of Knowledge*...; vol. 4" followed by the reference to the German standard edition, i.e. ECW 5. Cf. E. Cassirer, *The Problem of Knowledge*...; vol. 4, p. 1 (ECW 5, p. 1): "Even in myth and religion all that is distinctive of man is associated with the miracle of knowledge. This miracle reveals the nature of man and his likeliness to God, yet in it man also realizes, in the deepest and most painful way, the very limitations of his nature. [...]. To this religious pessimism the Greeks were the first to take a definitely opposite view. It was a decisive affirmation that knowledge is certainly possible to man. There was no longer that sense of a "fall" of man and of an estrangement from the ultimate ground of things, but on the contrary a conviction that knowledge is the one power which can sustain and unite man forever with the ultimate being".

[5]E. Cassirer, *The Problem of Knowledge*...; vol. 4, pp. 47–48 (ECW 5, pp. 53–54): "The Greeks were able to unearth this hidden wealth because the idea of measure lay at the heart of their view of the world and of all their thinking. Restricted to no particular sphere and not exhausted by any special application, this idea represented the very essence of thinking and of being. To discover the "limits and the proportions of things" was the task of all knowledge. But the concept extended far beyond this purely intellectual achievement, since it was the core not only of all cosmic but also of the human order, and lay at the center of ethics as well as of logic".

which would aspire to constitute itself into a theory, but on the other hand, the resistance of knowledge to being shaped through geometry, generally attributed to the changing and measureless character of the objects which populate the sensible world and of which the representations are unfaithful by nature, manifests the possibility of a *perhaps partial* character of this geometrization, as if the theoretical ideal of knowledge did not limit itself to its representation through geometry and as if the first principles *of another order* were still required.[6] This ambivalence traverses the history of the relationships between philosophy and geometry, as we will see in the second point.

Secondly, from the epistemological standpoint, the extraordinary theoretical success represented by geometry does not only stem from its axiomatic mode of presentation, but also from its power in describing natural phenomena. From this point of view, the project of constituting geometrical physics at all cosmic scales, quite prior to its being put into axiomatic form since it begins with Anaxagoras and Democritus in the fifth century B.C. long before Euclid, appears to deliver proof of the possible constitutive role played by geometry in all fields of knowledge. Even though, as Cassirer has shown, modern physical science distinguishes itself from Greek science through its purely functional aspect—which profoundly unsettled the type of mathematical tools implemented[7]—this capital change in the nature of science only corroborates the philosophical option according to which an extension of the geometrical domain to the objects of the physical world is quite achievable, beyond the ideal mathematical objects which are geometrical figures. However, the ambivalence surrounding geometry in the explicitation of the first principles nevertheless remains. Indeed, it is completely permeated by the *theoretical* viewpoint inherited from the Ancients[8] that mathematical physics constituted itself during the seventeenth century around the figures of Galileo and Descartes, but it is *exclusively* the domain of physical nature that falls under the scope of this mathematical extension, whereas the other fields of knowledge of which the objects are not directly localizable in mathematical space and time remain inaccessible to it. There is therefore a kind

---

[6]The role attributed to Greek skepticism as well as to French skepticism as seen in the tradition of Montaigne, for instance, plays an eminent role for Cassirer in this conclusion; cf. E. Cassirer, *Das Erkenntnisproblem...*, Band 1, ECW 2, pp. 143–168.

[7]E. Cassirer, *Substance and Function*, p. 21 (ECW 6, p. 20): "In opposition to the logic of the generic concept, which, as we saw, represents the point of view and influence of the concept of substance, there now appears the logic of the mathematical concept of function. However, the field of application of this form of logic is not confined to mathematics alone. On the contrary, it extends over into the field of the knowledge of nature; for the concept of function constitutes the general schema and model according to which the modern concept of nature has been molded in its progressive historical development".

[8]E. Cassirer, *Das Erkenntnisproblem...*, Band 1, ECW 2, p. 27 for the particular role played by Plato in the philosophy of Nicholas of Cusa; more generally, on Plato's role during the Renaissance, cf. Ernst Cassirer, *Das Erkenntnisproblem...*, Band 1, ECW 2, p. 65sq. Cf. also E. Cassirer, *The Problem of Knowledge...*; vol. 4, pp. 1–2 (ECW 5, p. 2): "To Galileo the "new science" of dynamics, which he founded, meant first of all the decisive confirmation of what Plato had sought and demanded in his theory of ideas. It showed that the whole Being is pervaded through and through with mathematical law and thanks to that is really accessible to human knowledge".

of limitation assigned to the *theoretical* project originally belonging to geometry, as Cassirer stressed concerning the negative diagnosis Pascal made regarding the Cartesian system.[9]

Thirdly, from a philosophical standpoint, the huge success of the constitution of mathematical physics beginning in the seventeenth century provoked a profound renewal of the very conception of knowledge: the indissolubly physical and mathematical character of knowledge requires to define its nature in a different manner. It can no longer be question of an extension of the geometrical beyond its realm proper, particularly towards physics; quite to the contrary, it is this physical and mathematical mixture which represents the stemming point of knowledge: any concept of which we would want to abstractly present the content without its corresponding physical domain would now appear as a simple abstraction. Cassirer, following his teachers at Marburg, takes note of the full reversal of perspective specific to modern science and identifies Kant as the one who drew all its philosophical consequences. Kant indeed demonstrated that in order for each concept to correspond to a determined object, it is necessary to suppose the intervention of a judgment associating the concept and the object within a *localized* space-time,[10] failing which, Reason will follow its tendency towards globality and infinity, beyond any precise determination within mathematized space-time.[11] Now, for Kant, this space-time is only mathematizable because it is based on the evidence of the Euclidean intuition of space and time[12]: *Euclidian*

---

[9]E. Cassirer, *Das Erkenntnisproblem…*, Band 1, ECW 2, pp. 439–441: "But from an historical point of view, Pascal's doctrine in contrast constitutes a symptom of a general, inner flaw in the Cartesian system. The strict separation between Reason and Authority remains limited to the theoretical domain. (…). There is no any other moment in the history of modern philosophy where one cannot become more clearly aware of the conflict between the two paths, between the two directions of thinking and of the alternative that makes it necessary to choose between the two than in the system of Pascal".

[10]E. Cassirer, *Das Erkenntnisproblem…*, Band 2, ECW 3, p. 624: "That the concept of an object would only present itself to the "Pure Understanding" certainly entails no contradiction and, from a purely logical point of view, cannot be disputed nor refuted; but this freedom from contradiction will here, just like in the case of all ontological concepts, be paid at the price of a complete lack of determined content".

[11]Thus, what constituted for Cassirer the very originality of the modern project of knowledge, that is, the tendency to relate the mathematized usage of infinity to the category of subject without attributing it to the transcendental, a tendency which had been already discovered to be at work at the cusp of modernity with Nicholas of Cusa (1401–1464) became, for Kant, a project for the self-limitation of the powers of reason, likely to convert all which appears as an object of the world into a simple correlate of a function of knowledge. Cf. E. Cassirer, *Das Erkenntnisproblem…*, Band 2, ECW 3, p. 633: "Gradually, all the real properties of the "world" transmute into the methodical features of experience." For Cassirer's reflections regarding Nicholas of Cusa, cf. E. Cassirer, *Das Erkenntnisproblem…*, Band 1, ECW 2, pp. 17–50.

[12]It is not necessary here to mention the reflection by Kant (for example in the *Prolegomena to Any Future Metaphysics That Will Be Able to Present Itself as a Science*, § 13) regarding what remains non-geometrical in the perception of space, the experience of which being allowed for instance by non-congruent objects (symmetrical but not superimposable), because what is in question here is only the characterization of what is *mathematizable* in space (the case of non-congruent objects will finally be resolved geometrically by Legendre within an Euclidean framework; cf. Legendre, *Éléments de géométrie*, VI, proposition 25 (1794). It is only in the second edition of the *Critique*

geometry is therefore at the basis of the possibility of physico-mathematical knowl-edge such as it has developed, during modern times, from Galileo to Newton. Thus recapturing the capital role which it had always had since the Greeks, even if it was in a profoundly different scientific context, Euclidean geometry remained in the end for Kant that upon which any scientific determination of nature rested.[13]

Now that the key role played by geometry in the establishment of science and in its relationships with philosophy has been somewhat clarified, let us return to the main topic of this chapter regarding the transformations internal to geometry over the course of the second half of the nineteenth century and to the philosophical problem these transformations represented for Cassirer.

1.2   Brief Reminder Concerning the Transformations of Geometry During the Nineteenth Century

In order to understand Cassirer's position, it is necessary to begin by briefly exposing the problem represented by the synthesis of three relatively autonomous currents in the mathematical works of the nineteenth century: projective geometry, group theory and non-Euclidean geometries.

What is meant by "projective" geometry is a geometry in which is studied what remains invariant in the geometrical figures following projection from a focal point (from the horizon, for example). Such geometry enables to generate by means of projection various figures and to establish the rules for the transformation of these figures from the ones into the others. Also, in projective geometry, there is for instance no longer a strict distinction between a circle, an ellipse, and a hyperbole inasmuch as these figures are all susceptible to being generated from a same transformation rule. This type of geometry, known since Antiquity[14] but which truly flourished

---

of Pure Reason, when it is question of exposing not the metaphysical concept of space but its transcendental counterpart, that the question of the Euclidean character of space and its necessity is posed. Cf. F. Pierobon, *Kant et les mathématiques; la conception kantienne des mathématiques*, Vrin, Paris, 2003, p. 110.

[13] Kant, *Critique of pure Reason*, 2nd edition, Transcendental Aesthetics, § 2: "Geometry is a science that determines the properties of space synthetically and yet a priori. What, then, must be the presentation of space be in order for such cognition of space to be possible? Space must originally be an intuition. For from a mere concept one cannot obtain propositions that go beyond the concept; but we do obtain such propositions in geometry. This intuition must, however, be encountered us a priori, i.e. prior to any perception of an object; hence this intuition must be pure rather than empirical. For geometric propositions are one and all apodeictic, i.e. linked with the consciousness of their necessity—e.g., the proposition that space has only three dimensions. but propositions of that sort cannot be empirical judgements or judgments of experience; nor can they be inferred from judgments" (*Critique of Pure Reason*, translated by Werner S. Pluhar, introduction by Patricia Kitcher, Hackett Publishing, 1996).

[14] For example, Pappus's theorem (fourth century AD) can retrospectively be linked to projective geometry.

in the nineteenth century with Poncelet in 1822, has the characteristic feature of not respecting a certain number of Euclidean properties (such as the invariance of distances and of angles after displacement, enabling the comparison of figures, that is, their measurement) all the while respecting others (such as the invariance of points or the concurrent aspect of straight lines).[15] Projective geometry thus leads to distinguish between two types of properties in geometry: the "metric" properties based upon the invariance in the measurement of distances and the "projective" properties which are preserved by central or perspective projection. Once Cayley had shown in 1859 that it was possible to define metric relations in projective geometry, projective geometry appeared to be vaster than did Euclidean geometry which formed only a part of it.[16] From an axiomatic standpoint, projective geometry was based on fewer axioms than was Euclidean geometry and, in particular, it did not employ Euclid's parallel postulate, of which the status had always been considered to be apart from the other axioms, due to the fact that it uses the notion of indefinite prolongation which was *conceived* and not *perceived*.[17] It is for this reason that mathematical research on the parallel postulate, independently from projective geometry, had led a certain number of mathematicians over the course of the nineteenth century[18] to renounce the Euclidean formulation of this axiom and to propose others, thereby producing other geometries, deemed "non Euclidean". Such research fundamentally questioned not only the Euclidean primacy of the description of space but also the very status of the axiomatic method itself, inasmuch as the axioms could no longer draw the descriptive power with which they were until then endowed from their relationship with external reality. Projective geometry, however, would enable to accommodate, within a same theoretical set, Euclidean and non-Euclidean geometries.

The necessity of classifying geometries had progressively led mathematicians to see in the notion of invariant by transformation the concept enabling to perform such classification.[19] It is starting with this question that group theory began to play an eminent role in geometry, when the notion of transformation was identified with

---

[15]In *Substance and Function*, p. 91 (ECW 6, p. 96): "Constancy and change thus appear as thoroughly correlative moments, definable only through each other. The geometrical "concept" gains its identical and determinate meaning only by indicating the definite group of changes with reference to which it is conceived. The permanence here in question denotes no absolute property of given objects, but is valid only relative to a certain intellectual operation, chosen as a system of reference".

[16]As Cassirer remarks in *Substance and Function*, p. 88 (ECW 6, p. 93): "In this connection, projective geometry has with justice been said to be the universal "a priori" science of space, which is to be placed beside arithmetic in deductive rigor and purity. Space is here deduced merely in its most general form as th "possibility of coexistence" in general while no decision is made concerning its special axiomatic structure, in particular concerning the validity of the axioms of parallels".

[17]In Euclid's text, the fifth axiom is indeed formulated: "That, if a straight line falling on two straight lines makes the interior angles on the same side less than two right angles, the two straight lines, if produced indefinitely, meet on that side on which are the angles less than the two right angles". (*The Thirteen Books of Euclid's Elements*, transl. T. L. Heath, University Press, Cambridge, 1908).

[18]Gauss, Riemann, Lobatchevski, Bolyai, Beltrami.

[19]Cf. Felix Klein, "A comparative Review of recent Researches in Geometry" (1872), English translation by Dr. M. W. Haskell and transcribed by N. C. Rughoonauth, *Bull. New York Math. Soc.* 2, (1892–1893), online at: http://arxiv.org/abs/0807.3161.

the notion of group.[20] Group theory, upon which work began around 1830, had for first objective the study of permutations of a finite number of objects. However, its field of relevance will have proven to be much broader.[21] It is Felix Klein who, beginning in 1872, made the connection between geometries and group theories,[22] in a major article called, in the mathematical tradition, the "Erlangen program".[23] This article established the nature of geometry in its full generality by reversing the order of priorities between geometry and the transformation group, which became the primary object of study. This led to two important results: on the one hand, Klein demonstrated that projective geometry was independent from the Euclidean parallel postulate; on the other hand, he established the projective nature of non-Euclidean geometries by showing that they were all particular cases of the general metrics developed by projective geometry. Thus, projective geometry enabled to offer a framework for the study of all geometries, be they Euclidean or non-Euclidean, as well as for their mutual relations.

Geometry had thus radically changed over the course of the nineteenth century. The geometry which preceded the synthesis operated by Klein in 1872 appeared

[20]Cf. Hans Wussing, *The Genesis of the Abstract Group Concept: a Contribution to the History of the Origin of Abstract Group Theory*, MIT Press, Cambridge, 1984, p. 198: "It is this very linking of two advances—the introduction of groups of geometric motions and of the question of their generators—that has produced the advance that we encounter in Klein's studies of groups of isometries of regular polyedra. This advance enabled Klein not to apply to geometry the fundamental principle of permutation theory but also to work out the concept of a (discrete) group of transformations. The recognition and clarification of that concept's potential to fuse geometry, algebra, and the theory of functions set in motion a far-reaching development".

[21]E. Cassirer, *The Problem of Knowledge...*; vol. 4, pp. 42–43 (ECW 5, pp. 48–49): "Since its early development by Cauchy, Lagrange and Galois the range of these applications has broadened continuously and extended to the most various fields of mathematics. The group concept is not restricted either to mathematics or geometry or to number or dimension. In it we rise above any considerations of special elements of mathematical thought to a theory of *operations*. We are no longer concerned with the special content but rather with the very procedure of mathematics itself, for we have entered the realm of pure, "intellectual mathematics"".

[22]F. Klein, "A comparative Review of Recent Researches in Geometry" (1872), § 1: "The most essential idea required in the following discussion is that of a group of space-transformations. The combination of any number of transformations of space is always equivalent to a single transformation. If now a given system of transformations had the property that any transformation obtained by combining any transformations of the system belongs to that system, it shall be called a group of transformations. [...]. Now there are space-transformations by which the geometric properties of configurations in space remain entirely unchanged. For geometric properties are, from their very idea, independent of the position occupied in space by the configuration in question, of its absolute magnitude, and finally of the sense in which parts are arranged. [...]. The totality of all these transformations we designate as the principal group of space-transformations; geometric properties are not changed by the transformations of the principal group. And conversely, geometric properties are characterized by their remaining invariant under the transformations of the principal group. [...]. This is the general problem, and it comprehends not alone ordinary geometry, boatels and in particular the more recent geometrical theories which we propose to discuss, and the different methods of treating manifoldness of n dimensions.

[23]E. Cassirer, *The Philosophy of Symbolic Forms*, vol. 3, p. 352 (ECW 13, pp. 405–406): "The theory of substitutions takes its place side by side with the theory of number developed in elementary arithmetic, and moreover it develops that the basic theorems of elementary arithmetic can be strictly

henceforth as an *empirical* geometry which accepted space at face value as it was *perceived*, without an awareness that such geometry limited its field of relevance to a specific transformation group; it had now become a *conceptual* science by making space into a correlate of the transformations operated on groups without involving, at least directly, the question of the perception of space. From the epistemological standpoint, this represented a veritable revolution; but how could this be assimilated when the reference epistemological framework was still that of the *Critique of Pure Reason* which was, as we mentioned, founded on the Euclidean character of geometry? Such was the problem to which Cassirer was confronted.

1.3   The Philosophical Problem Posed by the Renewal of Geometry

From a philosophical standpoint, the issue of the transformations of geometry over the course of the nineteenth century had mainly taken the form of a debate regarding the status to ascribe to non-Euclidean geometries. Even though, epistemologically, this point was but a part of the problem represented by the renewal of geometry, it was of course essential since it directly concerned the conception to have regarding the axioms of geometry as an axiomatic system of the measurement of distances. Let us first begin by citing the manner in which the problem of non-Euclidean geometries presented itself to Cassirer in one of the rare occasions where he returned to the philosophical issues with which he was confronted during his student years[24]:

> "For the sake of clearness and brevity, I begin here with a personal reminiscence. I remember very well the great difficulties I felt, when I began to study the different systems of non-Euclidean geometry. At this time I was a student of Kantian philosophy. I had read the *Critique of Pure Reason* with the deepest interest and a real enthusiasm. I had no doubts that the Kantian thesis – the thesis of the empirical reality and the transcendental ideality of space – contained the clue to the solution of this problem. But how was this thesis to be reconciled with the progress of geometrical thought, made in the nineteenth century? As for myself, I was quite unable to find any flaw, any vulnerable point in the new geometrical systems. Even from a philosophical and epistemological point of view they opened a much larger perspective; they were pregnant with new problems and new promises. [...]. I found the escape from the dilemma when I happened to study a short article of Felix Klein titled "Vergleichende

deduced only on the basis of this theory. And from here the road leads to what has been said to be "perhaps the most characteristic concept in nineteenth-century mathematics" [note: Hermann Weyl, "Philosophie der Mathematik und Naturwissenschaft", *Handbuch der Philosophie*, Munich & Berlin 1927, Pt II A, p. 23]. For investigations of groups of letter substitutions give rise to the general concept of group of operations and the new discipline of a theory of groups. Not only was the group theory an important addition to the system of mathematics, but it soon became increasingly plain that this was a new, far-reaching component of mathematical thought itself. Felix Klein's famous 'Erlanger Programm' shows how it changes the inner form of geometry. Geometry now becomes subordinated to the theory of invariants as a special case".

[24]"Reflections on the Concept of Group and the Theory of Perception", translated in D. P. Verene, *Symbol, Myth and Culture; Essays and Lectures of Ernst Cassirer 1935–1945*, (1979), pp. 277–279 (ECN 8, pp. 187–188).

Betrachtungen über neuere geometrische Forschungen" ("Comparative Studies into Recent Geometrical Investigations") – an article that, later on, was usually quoted as the so-called Erlanger Programm [1872] because it was the inaugural address of Klein when he took his chair of professor of mathematics in Erlangen. [...]. After having studied Klein's Erlanger Programm I began to see this problem in an entirely new light. What we have learned from the non-Euclidean geometries, said Klein, is that geometry is not as simple a thing as it was supposed to be in former times. We must give up our traditional view of geometry; we must seek for a new and deeper insight into the method and character of geometrical thought. This new insight was won, in this article of Klein, by the introduction of a new concept: the concept of group. The generalization of geometry, he said, leads to the following problem: "Given a multiplicity (Mannigfaltigkeit) and a group of transformations referring to the former; the problem is to study the elements of multiplicity with regard to those properties which are not affected by the transformations of the group."

Regarding the issue of non-Euclidean geometries, it had been addressed and resolved by Felix Klein one year prior to the "Erlangen program" when he demonstrated how non-Euclidean geometries could be transposed into the framework of Euclidean geometry.[25] Yet, the solution offered by Klein still left open the philosophical problem concerning the very notion of truth.

Cassirer referred on numerous occasions to the scandal that the emergence of non-Euclidean geometries continued to represent for certain philosophers who taught during the time he was still a student.[26] And it is easy to understand why: what was at stake in the multiplication of geometries was not the technical problem it posed regarding mathematics since that had been resolved by means of the well-founded concept of group, but rather the danger of a *complete ruin of the conception of rationality as it was conceived of since the common origin of both geometry and philosophy.* Indeed, as we have seen above, the invariance of geometry through time, its systematic coherence reflected in its axiomatic basis, its exclusive usage of demonstration, its link with the possibility of describing the physical world, and the thoroughly theoretical requirement shared with philosophy, all of this was countered by the non-Euclidean geometrical systems about which it was reasonable to wonder

---

[25] E. Cassirer, *The Problem of Knowledge...*; vol. 4, p. 25 (ECW 5, p. 28): "In 1871, Felix Klein showed that the entire system thereof can be derived from Euclidean geometry, a fact that makes illusory every preference for one over the other on the score of absolute worth and proves that they share the same fate in respect to "truth", since any contradiction in one is inevitably attended by similar contradictions in the others".

[26] This was, for instance, the case with very famous philosophers such as Lotze who declared (ECN 8, p. 186): "[...] I plainy say that the whole of this speculation seems to me one huge coherent error [Hermann Lotze, *Metaphysics*, Oxford, Clarendon Press, 1887, I bk. 2, ch. 2, p. 276]." (quoted in Cassirer "Reflections on the Concept of Group and the Theory of Perception", translated in D. P. Verene, *Symbol, Myth and Culture; Essays and Lectures of Ernst Cassirer 1935–1945,* (1979), p. 277) or Wundt who stated, in the strictest of Kantian orthodoxies (quoted in E. Cassirer, *The Problem of Knowledge...; vol. 4, p. 28 (ECW 5, p. 32)): ""This question is on a par with that raised more than once in the older ontology: wether or not the real world is the best among all possible worlds. No one since Kant has hesitated to reply that the real world is the only one that exists, and that nothing at all can be said about the nature of worlds that do not exist" (W. Wundt, "der mathematische Raumbegriff", *Logik*, I, p. 496, 2. Aufl. Stuttgart, Ferdinand Enke, 1893)".

upon what foundations they rested because the *real* character of their descriptions was no longer objectively ensured. As stated by Cassirer[27]:

> If geometry owes its certainty to pure reason, how strange that this reason can arrive at entirely different and wholly incongruous systems, while claiming equal truth for each! Does this not call in question the infallibility of reason itself? Does not reason in its very essence become contrary and ambiguous? To recognize a plurality of geometries seemed to mean renouncing the unity of reason, which is its intrinsic and distinguishing feature.

Indeed, up till then, it had been considered that the truth of a proposition could be founded upon two sources: either its *evidence* or its *demonstration* on the basis of other propositions considered to be evident. It is this double source which enabled to extend the legality of ideal geometry towards the space perceived through the senses which was also considered, up till then, as being the space of physics. To sever this link, it appeared, would be to imperil the *rational project of knowledge about nature.* Furthermore, from the standpoint of the very nature of geometry, the danger that these systems represented for rationality had a name which philosophy, since its Greek origins, had always sought to denounce and oppose: *relativism,* which dangerously threatened the principle of non-contradiction since, for a same proposition, the difference between affirmation and negation was no longer a matter of truth but one of arbitrary choice. In the case at hand, such relativism touched the axioms of geometry itself because Euclidean and non-Euclidean axioms were mutually incompatible. So, then what would a theory such as projective geometry be worth if it was capable of encompassing geometries which were so manifestly contradictory?

It was therefore necessary to completely review the very notions of space, of geometry, and of axiom, but also those of truth and of reality in order to make sense of the new research developed by Klein. It is in this way that mathematical research entailed specifically philosophical work, work to which Cassirer would devote himself.

2   Epistemological and Philosophical Consequences

There are three of them.

2.1   *Truth and Abstraction*

Since the geometry of ancient Greece, figures were deemed to be *abstract.* What was meant by this? In short, it meant that the empirically drawn figures only provided concrete assistance for representing invariant and ideal figures that were not subject to change. Abstraction thus went hand in hand with invariance with respect to the sensible realm. This theory of abstraction thus rested upon the idea that the distance

---

[27] E. Cassirer, *The Problem of Knowledge…*; vol. 4, p. 24 (ECW 5, pp. 26–27).

which separated the sensible realm from the intelligible one was also what separated empirical knowledge from *a priori* knowledge. However, the point of view developed by Klein put an end to this manner of seeing things.

Firstly, it was no longer with respect to the sensible realm that geometry could define its "abstraction", the relationship with perceived space having definitely been severed. Abstraction as seen by Klein was defined in a manner *internal* to geometry because it was possible to define a hierarchy between groups, inducing a hierarchy between geometries: any extension from the group of Euclidean geometry led both to the loss of certain properties considered to that point to be invariant, as well as to the maintenance of some others which, by resisting the extension of the group, revealed their more profound invariance.[28] Then, from the standpoint of the nature of knowledge, the continuity established by ancient theory between sensible knowledge and intelligible knowledge was no longer relevant because the relationship to the sensible realm had been definitely discarded from the definition of invariant properties. For Cassirer, the relationship between knowledge and the sensible realm would be reestablished by means of a *novel usage of the concept of group*, oriented towards matter, be it through physics or through the theory of perception.[29] Finally, the most important point was the fact that equally true geometrical systems came to be defined *all the while avoiding relativism*: one geometry was as true as another, once account taken of the reference system within which its invariants were defined.[30] By avoiding relativism, Klein's theory would have founded, for Cassirer, a theory of the *relativity* of invariants, according to the reference system within which they were inscribed. As we have seen, no geometry is truer than another and only the system of geometry in its globality must be conceived as ultimately true, even if such a system remains to be devised. So there lied the seeds of a new conception of the *a priori* where the relativity of the reference systems made possible the search for evermore fundamental invariants. Cassirer would later remember this at the moment of devising the notion of a "symbolic form", conceived as a global cultural system diversifying into specific invariants such as language, myth, and science.

[28] E. Cassirer, *The Philosophy of Symbolic Forms*, vol. 3, pp. 352–353, (ECW 13, p. 406): "What links the various geometries is that each of them considers certain basic properties of spatial forms, which prove invariant in relation to certain transformations; what distinguishes them is the fact that each one of these geometries is characterized by a particular transformation group. [cf. F. Klein, "Erlanger Programm; Vergleichende Betrachtungen über neuere geometrische Forschungen", *Mathematische Annalen*, 43, 1893 pp. 63–100]".

[29] In his last manuscript (April 1945), Cassirer described the two directions in question by mentioning on the one hand Hermann Weyl's book *Gruppentheorie und Quantenmechanik* [*The Theory of Groups and Quantum Mechanics*] and, on the other hand, *Gestalt Theory* as a theory which sought to establish new foundations for a theory of perception, particularly with respect to "perceptual constancy", using the concept of invariant by transformation inherited from group theory. For these two points, see below.

[30] E. Cassirer, *The Problem of Knowledge...*; vol. 4, p. 33, (ECW 5, p. 38): "In passing from one geometry to another the peculiar change in meaning is constantly seen".

2.2   Truth, Axiomatics, and Language

The scandal of non-Euclidean geometries came from the perturbation they caused with respect to the definition of truth: up to that point, the truth of geometrical axioms came from their intuitive character, whereas they henceforth appeared to depend upon empirical content.[31] This situation having been judged to be damageable to the rational theory which philosophy aspired to be, for Cassirer, it was necessary to start by reestablishing the very notion of truth in its rightfully *a priori* capacity. Contrarily to many epistemologists of his time who, when confronted with the same problem, considered that the rational status of geometry depended upon its *logical* content alone,[32] Cassirer wished to preserve the link geometry enjoyed with *physics*, a link which he considered to be constitutive of modern rationality since the advent of mathematical physics in the seventeenth century.[33] It is through this problematic configuration that Cassirer saw in Hilbert's "implicit definition" theory, as it appeared in his 1899 work *The Foundations of Geometry*, the means by which to preserve the rational character of the synthesis between geometry and physics.[34]

For Hilbert, it is not the isolated and intuitively conceived axiom, but rather groups of axioms which determine, by the reciprocal relationships they induce, the meaning of the concepts which are contained within them. So, it is only from the three undefined primitive notions (the point, the straight line, and the plane) that five groups of axioms[35] enable to construct new objects and their reciprocal relations. In this perspective, the axioms are in fact "implicit" definitions, in the sense that they indicate

---

[31] E. Cassirer, *The Problem of Knowledge...;* vol. 4, p. 22, (ECW 5, p. 24): "The whole character of mathematics appeared radically changed by this view, and axioms that had been regarded for centuries at the supreme example of eternal truth now seemed to belong to an entirely different kind of knowledge. In the words of Leibniz, the 'eternal verities' had apparently become merely 'truths of fact'. It was obvious that no simple mathematical question had been raised here; the whole problem of the truth of mathematics, even of the meaning of truth itself, was placed in an entirely new light.".

[32] It was the same reason for which the Vienna circle defended the *logicist* thesis according to which the properly conceptual content of geometry could be fully reduced to logical propositions, its experimental content being limited to empirical facts.

[33] E. Cassirer, "Kant und die moderne Mathematik", *Kant-Studien*, 12, 1907, p. 46 (ECW 9, pp. 79–80): "Even if [experience] is unable to ground and legitimate mathematical concepts by itself alone, it can contribute to determine them more closely and to make a selection between various possible principles that we could consider the apex of our deductions with an equal logical right. If we were to deny experience its function of selection, we would take the risk of depriving the fundamental concepts of geometry of all explicit sense; we would have then no means to make any distinction at all between various complex structures, in so far as are fulfilled all the conditions that we have registered in the axioms".

[34] E. Cassirer, "Kant und die moderne Mathematik", *Kant-Studien*, 12, [1907], p. 27, (ECW 9, p. 61): "Thus from now on Geometry doesn't possess, compared to the general logical axioms, independent and unprovable theorems: what is usually called the 'geometric Axioms' are rather only implicit definitions".

[35] These five groups comprise the axioms of incidence, order, congruence, parallels and continuity.

only how to conceive of the types of relation (for instance symmetrical or transitive) between concepts.[36] The question that must then be resolved is that concerning whether the axioms can operate this determination of content without ambiguity. Hilbert admits and even defends the ambiguous character of axioms,[37] as a normal state of their usage in mathematics as opposed to formal usage as such which he defines as a univocal play of signs devoid of meaning. This ambiguity specific to mathematical usage does not however imply limitation to intuition, but only to a certain choice-dependent usage. From this point of view, it is the *usage* previous to any intuition of axioms which carries the possibility for the elaboration of meaning, one which is fully univocal in the sole case of formal axiomatics where there is strict adherence to only the rules generating the signs. Also, the philosophical problem posed by non-Euclidean geometry precipitates, in what concerns the very axioms of geometry, the end of a theory of *resemblance* between the sign and the designated object and heralds the beginning of a theory of *meaning* internally conceived, in a manner fully disjoined from the question of the relation of meaning with an externality via an interpretive function. The Hilbertian approach to axiomatics thus rehabilitates that which constitutes the very essence of the *linguistic construction of meaning* inasmuch as it is by mutual convergence that the meaning of words establish themselves and not through the intuition of a reality external to language. This parallel drawn between the axiomatic approach and a certain type of linguistic undertaking calls for two remarks that will prove crucial with respect to Cassirer's later philosophy.

Firstly, it is undoubtedly with the introduction of the Hilbertian notion of "implicit definition" in the axiomatic method that the theme of meaning and of language appears in the philosophical works of Cassirer, as an internal modification to be applied to the notion of truth. It is therefore indeed for reasons pertaining to the *internal transformations of geometry* and to its axiomatic foundation that Cassirer comes to relinquish the intuitive conception of truth and that he arrives towards the

---

[36]E. Cassirer, *The Problem of Knowledge...;* vol. 4, p. 26 (ECW 5, p. 29): "The single elements receive their roles, and hence their significance, only as they fit together into a connected system; thus they are defined through one another, not independently of one another. Hilbert has expressed and clarified this feature of mathematical thinking with the greatest precision in his theory of "implicit definition". It is clear that for this theory any given geometry can be from the first nothing but a certain system of order and relations, whose character is determined by principles governing the relationships, and not by the intrinsic nature of the figures entering into it. The same system can acquire as many and as highly varied elements as one wishes without in any respect losing its identity. Hence the points, straight lines, and planes of Euclidean geometry can be replaced in an endless number of ways by other and entirely different objects without the least change in the content and truth of the corresponding theorems".

[37]For example, when Hilbert responds to Frege: "Every axiom contributes something to the definition, and hence every new axiom changes the concept. A 'point' in Euclidean, non-Euclidean, Archimedean and non-Archimedean geometry is something different in each case. [...]. In thinking of my points I think of some system of things, e.g. the system: love, law, chimney-sweep... and then assume all my axioms as relations between these things, then my propositions, e.g. Pythagoras' theorem, are also valid for these things." G. Frege, *Philosophical and Mathematical Correspondence*, ed. By G. Gabriel, H. Hermes, F. Kambartel, C. Thiel, A. Veraart, trans. By H. Kaal, The University of Chicago Press, 1980, p. 40.

theme of *meaning*. Truth then appears as being conceived in function of two features, as Cassirer points out[38]: on the one hand, it is no longer interpreted as a definite acquisition but as an *unending process*,[39] in the manner in which Klein's works in geometry have progressively uncovered increasingly profound geometric invariants; on the other hand, it manifests in the form of *configurations* which are independent from one another but which may nevertheless communicate when the conceptual means for such joining becomes possible, once again as Klein demonstrated in geometry.

Secondly, we also understand how this still embryonary point of view nevertheless exceeds the epistemological point of view, *stricto sensu*: rationality is no longer defined solely in reference to a mathematico-physical science conceived as its paradigm but it involves typically *semiotic* features which pertain to a whole other form of rationality and which do not have axiomatic intuition for principle. These themes will form the basis of the project defended by Cassirer in *The Philosophy of Symbolic Forms* from 1923 onwards, a project which must be interpreted with respect to the epistemological framework appertaining to the evolution of geometry as it took place between 1872 (date of Klein's Erlangen Program) and 1899 (date of Hilbert's *The Foundations of Geometry*).

2.3   Truth, Axiomatic Coherence, and Geometric Choice

The research work produced by Klein and Hilbert therefore led to distinguish two questions in geometry: axiomatic coherence and the relation of the axiomatic method to physical space.

The issue of axiomatic coherence would be explored through the criterion of *logical* non-contradiction, as Hilbert had begun to do already with his text of 1899. We must immediately remark that Cassirer did not further follow, after 1899, Hilbert's

---

[38] E. Cassirer, *The Problem of Knowledge*...; vol. 4, p. 25 (ECW 5, p. 28): "In his essay *On So-called Non-Euclidean Geometry*, published in 1871, Felix Klein showed that the entire system thereof can be derived from Euclidean geometry, a fact that makes illusory every preference for one over the other on the score of absolute worth and proves that they share the same fate in respect to "truth", since any contradiction in one is inevitably attended by similar contradictions in the others. His proof was completed and made still more rigorous by Hilbert, who showed in his *Foundations of Geometry* that the theorems of the various systems are reflected not only in one another but in the purely analytical theory of real numbers, so that every contradiction in them must also appear in this theory".

[39] E. Cassirer, *The Philosophy of Symbolic Forms*, vol. 3, pp. 475–476 (ECW 13, p. 555): "For here it is not a matter of disclosing the ultimate, absolute elements of reality, in the contemplation of which thought may rest as it were, but of a never-ending process through which the relatively necessary takes the place of the relatively accidental and the relatively invariable that of a relatively variable. We can never claim that this process has attained to the ultimate invariants of experience, which would then replace the immutable facticity of "things"; we can never claim to grasp these invariants with our hands so to speak Rather, the possibility must always be held open that a new synthesis will instate itself and that the universal constants, in terms of which we have signalized the "nature" of certain large realms of physical objects, will come close together and prove themselves to be special cases of an overarching lawfulness".

axiomatic reflection; from this standpoint, all the latter's work regarding the *formalization* of axiomatics in the 1920s and what has since been called "Hilbert's Program" [40] barely appears in his works.[41] The constitution of mathematical logic into an autonomous mathematical discipline (computability theory, model theory) therefore completely eluded Cassirer, most likely because on the one hand, this new discipline would have diverted him too much from the relationship between geometry and physics which he deemed to be consubstantial with modern rationality and, on the other hand, because he opposed Hilbert's "formalist" stance in the dispute concerning the foundations of mathematics,[42] having himself opted for an "intuitionist" standpoint such as the one defended by the mathematician Hermann Weyl.

The second question concerned the relation of the axiomatic method to physical space and, more specifically, the *choice* of a specific axiomatic system in order to account for physics.

It was of course this notion of choice which had appeared to be so contrary to the position defended by all partisans of the intuitive theory of axiomatic systems and which was now necessary to justify. By basing himself on Poincaré's epistemology, specifically as illustrated in the latter's 1902 work *La science et l'hypothèse*[43] which

---

[40] What is traditionally called in retrospect "Hilbert's programme" (see for example Georg Kreisel, "Hilbert's Programme", *Dialectica*, Volume 12, Issue 3–4, Dec. 1958, pp. 346–372) was developed during the 1920s, using an approach said to be "meta-mathematical" which exclusively concerned the form (as in *written form*) of mathematical statements and consisted in an attempt to demonstrate the internal coherence of the whole axiomatic approach by reducing the question of this coherence to the sole coherence of integers. Cassirer allows for (without adhering to) this specifically Hilbertian conception of the sign even if he does not follow the ulterior development of Hilbert's program as such concerning the internal coherence of formal axiomatics nor of the controversies that such a program sparked. Cassirer cites for example Hilbert's article of 1922 "Neubegründung der Mathematik" ["New Foundations of Mathematics"], *Gesammelte Abhandlungen*, Band III, Springer Verlag, 1970, pp. 156–177 in E. Cassirer, *The Philosophy of Symbolic Forms*, vol. 3, pp. 379–380 (ECW 13, p. 437): "In diametrical opposition to Frege and Dedekind, writes Hilbert in summing up his fundamental point of view, "I find the objects of the theory of numbers in the signs themselves, whose form we can recognize universally and surely, independently of place and time and of the special conditions attending the production of the signs as well as of insignificant differences in their elaboration. Here lies the firm philosophical orientation which I regard as requisite to the grounding of pure mathematics, as to all scientific thinking, and communication. 'In the beginning,' we may say here, 'was the *sign*'"."

[41] Cassirer's allusion to Hilbert in volume 3 of *The Philosophy of Symbolic Forms* pertains only to the status conferred by Hilbert to the sign in mathematics and which he compares to Leibniz. Cf. E. Cassirer, *The Philosophy of Symbolic Forms*, vol. 3, p. 379 (ECW 13, p. 437): "It is a critical authority of this sort that Hilbert strives to create in his theory of proof. Here the fundamental idea of Leibniz' "universal characteristic" is resumed and given pregnant and acute expression. The process of verification is shifted from the sphere of content to that of symbolic thinking. As precondition for the use of logical inferences and for the practice of logical operations, certain sensuous and intuitive characters must always be given to us. It is in them that our thinking first gains a sure guiding thread, which it must follow if it wishes to remain free from error".

[42] This expression designates the often heated debates between the years 1910 and 1930 opposing the advocates of three theses concerning the nature of the number with respect to a new arithmetics of infinity elaborated by Cantor: Frege's and Russell's logicism, Hilbert's formalism and Brouwer's and Weyl's intuitionism. I will return to this.

[43] Henri Poincaré, *Science and Hypothesis*, New York, Dover Publications, 1952.

represents a reflection regarding the usage of the notion of group by Felix Klein, Cassirer attempts to show that the notion of axiomatic choice does not throw the *a priori* character of geometry into question,[44] even if it modifies the relationship between axiomatic theory and experience.

For Cassirer, the multiplicity of the axiomatics of geometry is the very proof of their *a priori* character with respect to experience, because if geometry depended upon experience, it could only be unique.[45] The axioms of geometry are henceforth not to be considered as absolutes, but are rather part of a pool of propositions serving as guidelines to the elaboration of new propositions, within particular reference systems all having the same value.[46] And yet, as Cassirer remarked, a physical argument had been used in an attempt to reestablish the privilege conferred to Euclidean geometry, at least in physics. The argument goes as follows: even if there is no particular set of axiomatic propositions which *theoretically* has a privileged relation with experience, there is one which *pragmatically* does have one, Euclidean geometry, in that it enables the description of the physical world, contrarily to other geometries. If we subscribe to this argument, the Euclidean privilege will have only shifted from theory to practice because it would be this geometry and this one alone which, finally, would ensure the fecundity of physics. In fact, such an argument is no longer receivable since Einstein's constitution of the theory of relativity. Indeed, from the moment when a physical theory demonstrates that it is necessary to have recourse to non-Euclidean geometries in order to define its system of measurement—which is the case for Einsteinian relativity—it becomes clear that no geometry has a privileged relationship with the description of physical space, and that the latter is quite *neutral* regarding the question of the choice of its geometric description. This had several consequences, from the standpoint of the relation between geometry and physics.

First, how to justify the fact that any geometry, whichever it may be, can be said to relate to an adequate description of *reality* if geometry and physical *reality* pertain

---

[44]E. Cassirer, *The Problem of Knowledge…*; vol. 4, pp. 42–43 (ECW 5, pp. 48–49): "The group concept is not restricted either to mathematics or geometry or to number or dimension. In it we rise above any consideration of special elements of mathematical thought to a theory of *operations*. We are no longer concerned with the special content but rather with the very procedure of mathematics itself, for we have entered the realm of pure, "intellectual mathematics". Poincaré appears to have been the first thinker to draw the consequences of this for the problem of geometry, and the result was a categorical denial of geometrical empiricism. For if the theory of the group is what must be introduced into the definition of geometry, and if each geometry can be designated as a theory of invariants in respect to a certain group, then a pure a priori element has entered into the conceptual definitions".

[45]E. Cassirer, *The Problem of Knowledge…*; vol. 4, p. 45 (ECW 5, p. 52): "The autonomous nature of mathematical thought may now be recognized in full measure, and we may see convincing proof of this autonomy precisely in the possibility of setting up a plurality of entirely independent systems of axioms".

[46]E. Cassirer, *The Problem of Knowledge…*; vol. 4, p. 45 (ECW 5, p. 52): "[…] here the modern concept of axioms differs characteristically from the ancient. They are no longer assertions about contents that have absolute certainty, whether it be conceived as purely intuitive or rational. They are rather proposals of thought that make it ready for action—thought devices which must be so broadly and inclusively conceived as to be open to every concrete application that one wishes to make of them in knowledge".

to two strictly distinct realms? What is the connection which unites geometry and reality?

This adequation does not derive from a similitude between a concrete and an abstract point of view, as was believed during the time when Euclidean geometry was the only possible geometry: in fact, no particular geometry resembles experience because, whatever the form taken by experimental measurement when such geometry is used in physics, it does not have for object the nature *of* space, but only that of objects *within* a theoretically preconceived geometrical space.[47] Moreover, as Cassirer remarked following the physicist and philosopher of science Pierre Duhem (1861–1916) whose 1906 book *La théorie physique, son objet, sa structure*,[48] he cites extensively,[49] it is not possible to radically distinguish propositions which pertain to a fully "experimental" level from propositions belonging to an exclusively "theoretical" level: this distinction is an abstraction because any physical proposition already enlists a whole interpretive system given its relation to experience.[50]

With geometry and experience now being conceived as strictly separated, *the link which unites them must be of a whole different order than simple resemblance.* For Cassirer, who in this respect follows Hertz, Poincaré and Duhem, the relation between geometry and experience is of a *symbolic* nature, a relation of signification carried by signs.[51] Probably recalling Leibniz whom he often cites in this regard,[52] the

---

[47] E. Cassirer, *Einstein's Theory of Relativity*, pp. 430–431 (ECW 10, p. 94): "No measurement, as Poincaré objects with justice, is concerned with space itself, but always only with the empirically given and physical objects in space. No experiment therefore can teach us anything about the ideal structures, about the straight line and the circle, that pure geometry takes as a basis; what it gives us is always only knowledge of the relations of material things and processes".

[48] Pierre Duhem, *The Aim and Structure of Physical Theory*, New York, Atheneum, 1962; reprint Princeton, Princeton University Press, 1991.

[49] E. Cassirer, *The Problem of Knowledge...*; vol. 4, pp. 111–113 (ECW 5, p. 128–133).

[50] E. Cassirer, *The Problem of Knowledge...*; vol. 4, pp. 111–112 (ECW 5, p. 129): "Every judgement concerning an individual case, in so far as it purports to be a proposition in physics, already includes *a whole system* of physics. It is not true, therefore, that the science consists of two strata, as it were: simple observation and the results of measurement being in the one, theories built upon these in the other. Observation and measurement prior to all theory and independent of its assumptions, are impossible".

[51] E. Cassirer, *Einstein's Theory of Relativity*, pp. 432–433 (ECW 10, p. 96): "The reality which alone it [space] can express is not that of things, but that of laws and relations. And now we can ask, epistemologically, only one question: whether there can be established an exact relation and coordination between the symbols of non-Euclidean geometry and the empirical manifold of spatio-temporal "events." If physics answers this question affirmatively, then epistemology has no ground for answering it negatively. [...]. If it is seen thus, that the determination of this element as is done in Euclidean geometry, does not suffice for the mastery of certain problems of knowledge of nature then nothing can prevent us, from a methodological standpoint, from replacing it by another measure, in so far as the latter proves to be necessary and fruitful physically".

[52] Cf. E. Cassirer, *Substance and Function*, p. 43 (ECW 6, p. 44): "Leibniz whose entire thought was concentrated upon the idea of a "universal characteristic", clearly pointed out in opposition to the formalistic theories of his time, the fact which is essential here. The "basis" of the truth lies, as he says, never in the symbols but in the objective relations between ideas".

link, for Cassirer, operates not between individuated entities but between *orders*:[53] that of geometry deployed through signs and that of phenomena developed through measurements, themselves retranscribed using signs.[54] There are therefore no item-to-item resemblances between geometrical notions and physical notions, but only global coherences of signification of which it is possible to study the identities and differences. The relation between geometry and physics therefore solely pertains to the systematic organization of their significations, understood as the readable plane of expression[55] in which *series* and *configurations* between the meaning-conveying elements that are signs can be mutually compared.

Secondly, once the nature of the link between geometry and reality has been sketched out, the way in which the choice of a particular geometry is made remains to be clarified.

This choice immediately poses a problem given what has just been said regarding the strict separation between geometry and physics. Let us begin with an example: if we use the Euclidean reference framework to physically measure objects, any measurement will be Euclidean *by definition* and without there being any measured object capable of derogating from this or requiring, in itself, a change in the geometrical reference framework.[56] Hence the following problem arises: if the measurements which are made using a geometry can only be carried out in accordance with its axioms, where does it stem from that it may be necessary to *change* the geometry? And how then must the notion of geometrical *choice* be conceived?

The choice in question is not empirical, even if empirical causes may intervene,[57] but rather depends on an entirely philosophical *norm*, that of the quest for the greatest

---

[53] E. Cassirer, *The Problem of Knowledge...*; vol. 4, p. 114 (ECW 5, p. 133): "Here a particular symbol can never be set over against a particular object and compared in respect to its similarity. All that is required is that the order of the symbols be arranged so as to express the order of phenomena".

[54] E. Cassirer, *The Problem of Knowledge...*; vol. 4, p. 112 (ECW 5, p. 129): "The scientific statement would not describe the ocular, auditory, or tactile sensations of an individual observer in a particular physical laboratory but would state an objective fact of quite another sort—a fact that could be made known, of course, only if an appropriate language, a system of definite symbols, had been created for its communication".

[55] Cassirer borrows Kant's metaphor of the alphabetical readability of phenomena; E. Cassirer, *Einstein's Theory of Relativity*, p. 434 (ECW 10, p. 97): "What Kant says of the concepts of the understanding in general, that they only serve "to make letters out of phenomena so that we can read them as experiences" holds in particular of the concepts of space. They are only the letters, which we must make into words and propositions, if we would use them as expressions of the laws of experience".

[56] E. Cassirer, *Einstein's Theory of Relativity*, p. 434 (ECW 10, p. 97): "The particular geometrical truths or particular axioms, such as the principle of parallels, can never be compared with particular experiences, but we can always only compare with the whole of physical experience the whole of a definite system of axioms".

[57] E. Cassirer, *The Problem of Knowledge;* vol. 4, p. 44 (ECW 5, p. 50): "For example, we derive from experience the presupposition that it is possible for an object to move freely in all directions without any change in shape, and accordingly we eliminate from the possible geometries a definite group for which this assumption does not hold good. [...]. Experience is not the means of proving geometrical truths, though it can well serve as "occasional" cause, in furnishing a motive for and

unity of phenomenal fields, as Kant had already established using the term of "unconditionality" in the *Critique of Pure Reason* but of which it is necessary to preserve the spirit while renouncing the letter.[58] Cassirer seeks to show what this quest for unity means with respect to the specific case of the apparition of the theory of relativity which it attentively witnessed. Beside the fact that it constitutes a revolution in physics occurring around the same period given that it begins in 1905, it is also the first physical theory to have had a direct recourse to a non-Euclidean geometry. Cassirer endeavored to describe the true "miracle" which occurred therein when Einstein discovered that the metric properties of some non-Euclidean spaces could be directly employed in equations describing the gravitational field without involving new unknown forces.[59] This "miracle" had already been foreseen by Riemann in his 1854 habilitation defense,[60] as Cassirer recalls,[61] even if Riemann's formulations appear to him to remain overly tinged with realism with respect to space.[62] The question of the choice of geometry therefore shows the extent to which geometry and physics are closely intertwined in the elaboration of such a choice.

---

invitation to the development of certain aspects of these truths and the choice of one above the others".

[58] E. Cassirer, *Einstein's Theory of Relativity*, pp. 439–440 (ECW 10, p. 104): "The step beyond him [Kant], that we have now to make on the basis of the results of the general theory of relativity, consists in the insight that geometrical axioms and laws of other than Euclidean form can enter into this determination of the understanding, in which the empirical and physical world arises for us, and that the admission of such axioms not only does not destroy the unity of the world, i.e., the unity of our experiential concept of a total order of phenomena, but first truly grounds it from a new angle, since in this way the particular laws of nature, with which we have to calculate in space-time determination, are ultimately brought to the unity of a supreme principle, that of the universal postulate of relativity. The renunciation of intuitive simplicity in the picture of the world thus contains the guarantee of its greater intellectual and systematic completeness".

[59] E. Cassirer, *Einstein's Theory of Relativity*, p. 440 (ECW 10, p. 105): "A doctrine, which originally grew up merely in the immanent progress of pure mathematical speculation, in the ideal transformation of the hypotheses that lie at the basis of geometry, now serves directly as the form into which the laws of nature are poured. The same functions, that were previously established as expressing the metrical properties of non-Euclidean space, give the equations of the field of gravitation".

[60] Bernhard Riemann, "Über die Hypothesen, welche der Geometrie zu Grunde liegen", *Abhandlungen der Königlichen Gesellschaft der Wissenschaften zu Göttingen*, 1867, Band 13, pp. 133–150; English transl. "On the Hypotheses which lie at the Bases of Geometry", trans. William K. Clifford, *Nature*, vol. VIII, Nos. 183, 184, pp. 14–17, 36, 37.

[61] E. Cassirer, *Einstein's Theory of Relativity*, p. 440 (ECW 10, p. 105): "Riemann, in setting up his theory, referred to its future physical meaning in prophetic words of which one is often reminded in the discussion of the general theory of relativity".

[62] E. Cassirer, *Einstein's Theory of Relativity*, p. 440 (ECW 10, p. 106): "Instead of regarding "space" as a self-existent real, which must be explained and deduced from "binding forces" like other realities, we ask now rather whether the a priori function, the universal ideal relation, that we call "space" involves possible formulations and among them such as are proper to offer an exact and exhaustive account of certain physical relations, of certain "fields of force". The development of the general theory of relativity has answered this question in the affirmative; it has shown what appeared to Riemann as a geometrical hypothesis, as a mere possibility of thought, to be an organ for the knowledge of reality".

We can see how the evolution of geometry between 1872—year of Klein's works—and 1899—year of Hilbert's works—has played a capital role in the evolution of Cassirer's philosophical thought. Geometry plays a role of paradigm inasmuch as it appears as the field of knowledge *par excellence* in which our very conception of knowledge was completely modified: from a knowledge bound to principles deemed invariant, geometry has transformed its object by becoming the locus of evermore advanced research pertaining to the invariants themselves. Moreover, by radically transforming its primary object, space, the evolution of geometry has given rise, with Einstein, to the greatest revolution in physics since Newton, and has otherwise rekindled the link it had always enjoyed with physics since the ancient Greeks by founding the rationality of the philosophical concept of nature on new bases. Cassirer came to discover that the relationship between geometry and physics involved two notions: that of *norm*, since it is by seeking the most complete unity of knowledge that geometry had evolved towards ever-increasing unity, and that of *symbol*, since the link between geometry and physics is of a symbolic nature.

A question remains, however: why does it incur to geometry to blaze this new path towards universality and why use it as a stepping stone to make the leap and attempt to explore through human knowledge in general, as Cassirer did, the way in which norms and symbols are articulated? Geometry is indeed but a part of human activity and it would be possible, after all, to reject its exemplarity. But to leave it at that would be precisely to forget the philosophical lesson imparted by Klein's geometric revolution. Because geometry, in its new mode of existence, is not universal from the onset, but only tends towards universality: also, it is indeed due to its *intrinsic limitation* that geometry becomes a mode of access to universality. But under these conditions, all other activities, *as long as their specific limitations are definable*, become, in the same respect as geometry, legitimate modes of access to universality. It thus becomes possible to undertake a general philosophical program which would aim, by circumscribing the perimeter specific to any activity, to define its limitation and, in doing so, to assign it a role in the constitution of invariants and the search for universality. Such is precisely Cassirer's philosophical program.

Therein, geometry, such as it was re-elaborated by Klein, resounds with the German notion of culture from which it had been, until then, excluded. Because it is indeed impossible to not recognize the German character of the geometrical revolution initiated by Klein: it is in effect German culture following Herder which, by opposing the type of Roman universality first transmitted by the Medieval Catholic Church and later on by the French Revolution, had instead put forth the specific character of the national cultures and the universality of civilization as a simple horizon which cannot be accessed by any specific national culture. Thus, the German idea of universality was not itself conceived as that which is given once and for all—as was the case with Euclidean geometry—but as that which remains lacking and which remains to be constructed, as was the case for Klein according to whom geometric invariants accounted for specific geometries and for whom the search for increasingly universal geometric invariants remained a horizon to be constructed. This is why the search for invariants is endless: universalization is to be accomplished through the progressive differentiation of object domains and it is this differentiation itself which, by making possible the search for evermore universal invariants, extends limitlessly.[63] It is therefore the limited character of differentiation which becomes the very condition of an indefinite access to universality.

There is a pitfall to be avoided here in that there is no relativism in this undertaking: there is, of course, nothing to be identified as belonging to a "German geometry" in the way in which Klein produced geometric concepts, a "German geometry" which would have made geometry dependent upon contingent circumstances by eradicating its intrinsic necessity. But there is indeed, on the other hand, a way in which Germans of this time conceive of geometry, a geometry which remains, in its universality and its objectivity, as a horizon and which governs the way in which the very notion of result is to be conceived. It will be precisely the cultural disaster represented by Nazism, disaster from which Cassirer will greatly suffer, which will have endeavored to confound the spheres of nation and of concept by attempting to create a "German science" which will have ruined the very idea of a specific German culture, an idea which was very dear to Cassirer.

We thus understand what authorizes Cassirer to attempt to extend Klein's method beyond geometry and the natural sciences. Philosophically, this extension is only legitimate if it is motivated by a modification in the very apprehension of the notion of object of knowledge, because there would not be any sense in transforming the methods of knowledge without these transformations having an incidence on the very constitution of known objects, as geometry and physics have shown, each in their respective field of relevance. From this point of view, it is no longer possible to keep with the manner in which Kant conceived of the notion of object which consists in distinguishing three great types, according to whether they pertain to the fields of knowledge, of ethics, or of aesthetics. As we have seen, it is now as soon as we touch upon the field of knowledge that the notions of norm (usually reserved to ethics) and of symbol (generally pertaining to aesthetics), both become involved. If an extension of Klein's method is conceivable, it must therefore find support in a notion which, from the onset, pertains to all of the three domains which are knowledge, ethics, and aesthetics: this transversal notion, as we shall see, is the "symbolic form" which, through the variations of reference systems, captures, in specific invariants, something of the nature of universality.

Cassirer's philosophical project thus consists in attempting to answer the question of how norms and symbols are articulated within the gnoseological, ethical, and aesthetic activities of human culture. There lies a truly anthropological program which involves the study of the vectors specific to norms and symbols, vectors which

---

[63]Commenting on Herder's work, Cassirer points out the following in E. Cassirer, *Freiheit und Form; Studien zur Deutschen Geistesgeschichte*, ECW 7, pp. 124–125: "Thus restriction and "deprivation" do not mean a lack per se any longer, but rather the necessary condition for any individual perfection. A certain deprivation from knowledges, dispositions and virtues determines just as well as a "positive" quality the place from which its efficiency can and must come from in a particular nation and at a particular time. This very deprivation is immediately fruitful and inspiring as well for it is immediately distinctive and typical. The force of limitation does not simply resist the force of perfection but is rather another expression for it; it is only their combination that grants everything particular the determination of its being and its achievement".

account for the unity of all of human rationality, in the indefinite multiplicity of its activities. This indefinite character does not represent an obstacle regarding the unity of human knowledge because it is precisely with this multiplicity that the necessary limitation to the quest for universality can be measured.

Such is how it appears possible to answer this chapter's opening question regarding the reasons which lead Cassirer, initially a philosopher of exact and natural sciences, to broaden his field of investigation towards the more general issue of the construction of meaning, without limiting himself to its purely scientific construction.

Chapter 2
The Functional Viewpoint in Physics and Its Consequences on the Symbolic Aspect of Knowledge in General

Abstract Starting with Cassirer's distinction between Substantial and Functional knowledge as a means to account for the difference between Ancient and Medieval science on the one hand and Modern and Contemporary science on the other, this chapter focuses on the long march towards Functional physics in Modern times and its philosophical consequences. Counter intuitively, the process started with a semiotic revolution in philology during the Renaissance period and reached its climax with transcendental philosophy in Kant. Keeping track of this semiotic thread and interpreting the final steps in the history of physics as requiring a symbolic approach advocated by Poincaré and Duhem, Cassirer shows how a renewed transcendental philosophy based on the notion of form borrowed from Goethe should be considered as a resource for post-Kantian developments. Two case studies are under closer scrutiny in this chapter: Relativity theory and theoretical biology.

Keywords Renaissance philology · Grammatical versus functional construction of reality · Mathesis universalis · Logical and physical necessity · Transcendental role of symbolic processes · Kantian philosophy as a resource for post-Kantian developments: relativity theory and theoretical biology

The first chapter drew a number of conclusions concerning the role of geometry in the development of a physical theory of space. It had in particular been possible to show how Cassirer came to shift his point of view which, in the beginning, had been strictly epistemological, by broadening it towards the issue of a general theory of the construction of meaning. Endowed with this new perspective, attention should now be paid to the manner in which his philosophical reflection regarding science and its history, which had served as a starting point for his works, also progressively made its way into this new approach.

© Springer Nature Switzerland AG 2020
J. Lassègue, *Cassirer's Transformation: From a Transcendental to a Semiotic Philosophy of Forms*, Studies in Applied Philosophy, Epistemology and Rational Ethics 55, https://doi.org/10.1007/978-3-030-42905-8_2

Through his personal itinerary, Cassirer first sought to clarify the part played by the concept of function[1] in works pertaining to the general mathematization of the natural sciences in the seventeenth century. In the context of physics and chemistry, it is indeed the mathematical notion of function which he deemed capable of accounting for the profound transformations undergone during the modern era as opposed to Antiquity and the Middle Ages. Cassirer borrows the term "function" from Leibniz who promoted it in the mathematical field: in its Leibnizian sense, the mathematical function enables to regroup elements (whether they present themselves as quantities or not) in the form of a *series*, by making possible the establishment of internal relationships between these elements regardless of their number. This can also be found, for example, with Galileo who draws a serial relationship between the position of an object and its speed. Under certain conditions of homogeneity, this manner of treating elements makes possible the absolutely determined anticipation of their future evolution within a strictly defined mathematical framework, that of an exhaustively circumscribed space of possibilities. But the term 'function' also carries, with Leibniz, a broader meaning: the function is on the one hand part of what makes possible the regulated relationship between phenomena and on the other hand that which makes explicit the possibility of their reciprocal determination. This reciprocal determination is of a *harmonic* nature in that it possesses the capacity to allow for a variety of concordant viewpoints regarding the same phenomenon.[2] Cassirer makes his own use of the Leibnizian generalization of the concept of function but gives it another sense: he takes the evolution of this concept as it occurred within the Leibnizian itinerary and projects it onto the global evolution of the natural sciences since the inception of the modern era, which he considers to begin with Nicholas of Cusa during the early fifteenth century.[3] It is therefore the *progressive diffusion of the functional viewpoint throughout knowledge* which characterizes, for Cassirer,

---

[1]E. Cassirer, *Das Erkenntnisproblem…*, Band 1, ECW 2, p. 303: "The concept of function that we use as a logical model and a standard of measurement only shows that quantities mutually condition one another, leaving out the question which of the two elements we have to think as dependent or independent variables."

[2]E. Cassirer, *Das Erkenntnisproblem…*, Band 2, ECW 3, p. 156: "Leibniz takes as a starting point the concept of function of the new mathematics he is the first one to understand in its full generality and liberates it, since its first conception, from the domain of number and quantity. Armed with this new instrument of knowledge, he steps in the fundamental questions of philosophy. And it becomes clear to him that it is no rigid and dead instrument he has put hand on; rather, the more he moves forward, the more it gains content and inner richness. The abstract mathematical concept of function enlarges itself to the concept of harmony in ethics and metaphysics. […]. However, the harmony, in its fundamental meaning, means not only the relationship between body and soul nor the correspondence between the various individual substances and the series of their representations; originally, it is much more the accord between various and altering points of view presupposing one another which lets represent and interpret reality."

[3]E. Cassirer, *Das Erkenntnisproblem…*, Band 1, ECW 2, pp. 21–22: "Nicholas of Cusa was the first one to dare the proposition, standing far from the antique method of exhaustion, that the circle in its content and reality was nothing else than a polygon with infinitely many sides. The concept of 'limit' is raised to a positive meaning: it is due to the unlimited process of convergence that the limit itself can be grasped and apprehended in its certitude. This uncompletable process does not stand as a proof of an inner conceptual defect but rather as an evidence of its power and specificity:

the "modern" era. But this progressive diffusion does not, in his view and as would be expected from a historian and a philosopher of the exact and natural sciences, stem from mathematics and physics: it is from a reflection on language and style, and within the framework of what we would today call the "language sciences" that it operates, the natural sciences themselves only being impacted by the functional viewpoint relatively late, starting in the seventeenth century. It is this evolution which we will now describe.

From a specifically epistemological point of view which marks the starting point of his research, we have just seen that it is the advent of the functional standpoint which, for Cassirer, distinguishes the physics of the Ancients from that of the Moderns. By means of the opposition between substance and function, Cassirer would then attempt to justify the great historical divide within the natural sciences: the substantial theory of physical reality which was that of the Ancients, most particularly that of Aristotle and his medieval following, was progressively substituted with a *functional* conception of the same reality, between the fifteenth century Renaissance and the modern era of seventeenth century. The very notion of object of knowledge made accessible by this new approach was profoundly shaken. Far from limiting itself to objects conceived as substances, modern science managed to the contrary to view them abstractly by keeping with mathematized descriptions of their relations in ideal spaces, themselves conceived as susceptible of receiving mathematical determination in the form of numbers. To do this, geometry would, thanks to Descartes, endow itself with a space conceivable in terms of points and accessible by means of algebraically determined coordinates. The renewal of the conception of space during the seventeenth century and onwards was thus at the heart of the evolution of physics[4] which manifested once more its profound solidarity with the parallel evolution of mathematics. Later on, starting during the second half of the nineteenth century—as Cassirer himself had experienced during his years of study—it is once more a functional enrichment of geometry via group theory which would have profound consequences in physics, from the moment when the possibility of a geometrical description of physical space in non-Euclidian terms became conceivable, a process which culminates, as Cassirer noted, with the theory of general relativity in 1915. It is the same difference between substance and function which Cassirer uses as a tool for describing the revolution represented by quantum mechanics during the 1920s.[5] Because there too, a process of functional determination of the object is in effect:

---

only in an infinite object, an infinite process, can Reason reach the consciousness of its own power. [...] Now infinity is not a barrier anymore but rather the self-affirmation of Reason."

[4]E. Cassirer, "Mythic, Aesthetic and Theoretical Space", in *Ernst Cassirer; The Warburg Years (1919–1933)*, translated by S. G. Lofts with A. Calagno, Yale University Press, New Haven and London, 2013, p. 319 (ECW 17, p. 413): "If we attempt to reduce the epistemological development of the problem of space to a short formula, then it can be said that one of the basic tendencies of this development and one of its essential results is that out of the insight into the nature and properties of space, knowledge of the *primacy of the concept of order over the concept of* being is gained and increasingly established."

[5]The first result, attributed to Planck [1858–1947], dates back to December 1900, but will be theoretically justified only during the 1920s.

far from keeping with a substantial conception of the atom, on the contrary, it is by renouncing it that a point of view can be accessed according to which only the totality of observations grouped in the form of probabilistic laws give to the notion of atom a certain real consistency.[6] The difference between substance and function therefore appears for Cassirer as the fundamental epistemological tool enabling to account for the concept of *determination* and for its historical evolution: a substantial approach puts inner limitations to its own knowledge of nature for the very notion of substance is seen as imposing an external limit knowledge has no control of. Was progressively substituted to this approach a functional conception according to which the limitation is what experimental knowledge imposes upon itself in view of ensuring a *possible* determination of the object, thus eliminating any idea of an inaccessibility of the object to be determined. For Cassirer, the notion of function is thus at the center of the transcendental viewpoint conceived as possible knowledge and it is this schema of evolution which enables to go from a situation where knowledge is submitted to an external limit to a situation where it attributes a limit to itself, that is, where the subject provides itself with its own law, an evolution which defines, according to Cassirer, what it is to be understood as the "modern" era.[7]

From the point of view of his own progression, Cassirer was able to broaden towards the Renaissance on one hand and towards the Enlightenment on the other his strictly epistemological reflection which was at first centered upon his study of the seventeenth century. Such a broadening which enabled him to qualify the modern era from a general point of view was made possible by shifting his reflection, at first related to the sole natural sciences, towards the notion of *meaning* such as it was re-elaborated within the language sciences starting during the Renaissance. This is what we shall now address.

---

[6] E. Cassirer, *Determinism and Indeterminism in Modern Physics: Historical and Systematic Studies of the Problem of Causality*, pp. 132–133 (ECW 19, p. 160): "For it is precisely the stability of the atom which constitutes the basic problem of atomic physics: and how can this stability be accounted for theoretically when the presupposition of substantialism is discarded, when atoms are no longer regarded as rigid spheres? This, beyond doubt, represented a difficult task—but it is precisely the progressive solution of this problem, logically and physically, that is so significant for modern quantum theory."

[7] In philosophy, the birth of modernity remains, for Cassirer, marked by the Cartesian doctrine of the *cogito*. Cf. E. Cassirer, *The Individual and the Cosmos in Renaissance Philosophy*, Philadelphia, University of Pennsylvania Press, 1972, p. 123 (ECW 14, p. 143): "But now, new systematic questions suddenly emerge from the manifold and contradictory historical conditions. The conscious formulation of these questions is, of course, one of the latest products of Renaissance philosophy, first attained by Descartes—indeed, in a certain sense, first by Leibniz. Descartes discovered and defined the new 'Archimedian' point from which the conceptual world of scholastic philosophy could be raised out of its hinges. And thus we date the beginning of modern philosophy from Descartes' principle of the *Cogito*."

1 Sign and Nature in Modern Science

As we have just seen, what is meant by "modern thought" in the view proposed by
Cassirer can be divided into two historical periods of which the limits, obviously
imprecise, overlap: the first is that of the Renaissance and corresponds more or less
to the fifteenth and sixteenth centuries, while the second period corresponds to the
seventeenth and eighteenth centuries.[8] Each of these periods can be divided into two
subperiods, following a division by century.

1.1 From the Philological Renaissance
to the Physico-Mathematical Renaissance

In order to understand how, during the Renaissance, a new functional solidarity
between concepts and objects set itself into place, one must start by considering
Cassirer's remark according to which the physico-mathematical Renaissance was
indeed *preceded* by a philological Renaissance having for object the critique of the
Aristotelian conception of the role of language in the sciences.[9] Cassirer remarks
that Aristotle had constituted his ontological categories in implicit reference to the
Greek language: even if this correlation should not be exaggerated,[10] he nevertheless
remarks that the ontological category of substance derives from the linguistic category of substantive, the categories of quantity and of quality deriving from that of the

---

[8]The terms "modern" and "classical" may lead to confusion. "Modern thought" designates a movement which encompasses all manifestations of culture: the language sciences, the natural, theologi-
cal, esthetic, and political sciences. Cassirer describes its progressive extension over four centuries
based on his reflection on the role of language. During this period, seventeenth century science is
designated as "classical" because a very particular functional solidarity is established between a
specific modality of the constitution of objects of knowledge and some mathematical tools such
as the notion of function, a modality which continues to this day to imprint the very notion of
knowledge. It is the reason why this state of science is deemed "classical" in the sense that it plays
the role of an epistemological framework.

[9]E. Cassirer, *Das Erkenntnisproblem...*, Band 1, ECW 2, p. 112 where Cassirer declares concerning the three key figures of this first Renaissance, Lorenzo Valla [1407–1457], Lodovico Vives
[1492–1540] and Pierre de la Ramée [1515–1572]: "Right here, at the core of the humanistic
way of thinking, more and more signs were announcing the transition from the philological to the
mathematical-physical sciences." We owe to Lorenzo Valla the first purely philological critique of
the Scriptures (he maintains that the Acts of the Apostles had not been written by them) and of the
document of the Donation of Constantine which attributed the *Imperium* in the West to the Pope (it
is actually a fake written at the time of Pippin the Younger). To Lodovico Vives we owe the first
pedagogic critique of the place unduly taken by rhetoric in scholastic teaching and we also owe
to Petrus Ramus the first introduction of mathematical courses at the *Collège royal* (founded by
Francis I in 1530), which was to become the *Collège de France* in Paris.

[10]E. Cassirer, *The Philosophy of Symbolic Forms*, vol. 1, p. 126 (ECW 11, p. 62): "[...] it is surely an
exaggeration to say that Aristotle borrowed from language the fundamental distinctions underlying
his logical doctrines."

adjective, and those of time and place from the linguistic category of the adverb.[11] Medieval scholasticism[12] had, from this standpoint, roundly upheld Aristotle's line of thinking and had pursued research concerning the list of categories by borrowing from Latin grammar its general approach.

According to Cassirer, the shift in perspective characteristic of the modern era was made possible by a shift in attention from grammatical categories to the notion of *style*, that is, from a logic based on an *ontology* to a thoroughly *aesthetic* reflection, as demonstrated by the extraordinary changes in artistic forms during the Renaissance.[13] This stylistic transmutation was made possible owing to various semiotic operators, particularly those of *natural sentiment* and *humor.* The lyrical poetry of Petrarch plays a preponderant role here because it expresses a new approach to nature in which the ego is situated within nature rather than being separate from it.[14] As for humor, its role must be understood in opposition to the *seriousness* of scholasticism.[15] Humour takes distance from what seems to be an immutable reality by bringing to the fore what was previously occulted, that is, the active role played by the subject in the elaboration of new styles that institute new relationships to reality. In this context,

---

[11] E. Cassirer, *The Philosophy of Symbolic Forms*, vol. 1, p. (ECW 11, pp. 62–63): "In the category of substance we clearly discern the grammatical signification of the "substantive"; in quantity and quality, in the "when" and "where", we discern the signification of the adjective and of the adverbs of time and place—and above all the four categories of poieïn, and paskein, ekein and keisthai, seem to become fully transparent only when we consider them in reference to certain fundamental distinctions which the Greek language makes in its designation of verbs and verbal actions. Here logical and grammatical speculation seemed to be in thoroughgoing correspondence, to condition one another—and medieval philosophy, basing itself in Aristotle, clung to this correspondence between the two." However, this correlation must not be emphasized, as was noted in the previous footnote.

[12] By "scholasticism", we mean the period of intellectual renewal in Western Europe extending from the twelfth century to the Renaissance and which sought to articulate ancient philosophy and Christian revelation, generally based on the philosophy of Aristotle.

[13] E. Cassirer, *The Philosophy of Symbolic Forms*, vol. 1, p. 127 (ECW 11, p. 63): "However, when modern thinkers began to attack the Aristotelian logic, when they contested its right to be called "the" system of thought, the close alliance into which it had entered with language and universal grammar, proved to be one of its most vulnerable points. Assailing it at this point, Lorenzo Valla in Italy, Lodovico Vives in Spain, Petrus Ramus in France attempted to discredit the Scholastic-Aristotelian philosophy. At first the controversy was limited to the sphere of linguistic study: it was precisely the "philologists" of the Renaissance who, on the basis of their deepened understanding of language, demanded a new "theory of thought". They argued that the Scholastics had only seen only the outward, grammatical structure of language, while its real kernel, which is to be sought not in grammar but in stylistics, had remained closed to them. The great stylists of the Renaissance attacked syllogistics and its "barbarous" forms, not so much from the logical as from the aesthetic angle."

[14] E. Cassirer, *The Individual and the Cosmos in Renaissance Philosophy*, Philadelphia, University of Pennsylvania Press, 1972, pp. 143–144 (ECW 14, pp. 165–166): "The lyrical mood does not see in nature the opposite of psychical reality; rather it feels everywhere in nature the trace and echo of the soul; for Petrarch, landscape becomes the living mirror of the Ego. [...]. Nature is not sought and represented for its own sake; rather, its values lies in its service to modern man as a new means of expression for himself, for the liveness and the infinite polymorphism of his inner life."

[15] For Cassirer, it is particularly in literature that this function operates, be it in the works of Boiardo, Ariosto, Rabelais, Shakespeare, or Cervantes. But it is in England that it reaches its peak and

humor has a directly aesthetic impact, as Cassirer's mentor at Marburg, Hermann Cohen, had already noted.[16] As surprising as this may seem at first glance, humor had indeed made it possible to steer away from an implicitly *grammatical* apprehension of reality. But this role of humor in the aesthetic elaboration of novel styles produced an effect which may appear even more paradoxical: according to Cassirer, it paved the way for a *mathematical theorization of nature* through a rekindling of the interest in the Platonic concept of idea. There are two aspects to the Platonic concept of idea: the first, which is well known, falls directly under the perspective of the constitution of a natural science following the *Timaeus* or the *Philebus*, as was readily understood by Renaissance mathematicians and physicists.[17] The second, on the other hand, is more subtle: it concerns the paradoxically *aesthetic* aspect of the Platonic concept of idea and the capital role it continued to play over time in the development of a geometrical approach to nature.[18] Also, it is owing to the aesthetic aspect of the Platonic concept of idea that the notion of objectivity gradually gained acceptance during the Renaissance,[19] by bringing the notions of geometrical form and of law

---

Shaftesbury [1671–1713] represents, according to Cassirer, its first theoretician when he stresses that: "Thus humour is not directed against the seriousness of knowledge or against the dignity of religion; but simply against a mistaken seriousness and an arrogated dignity, against pedantry and bigotry. To the pedan, as to the zealot, freedom of thought is an abomination; for the former takes shelter from it behind the dignity of knowledge, the latter behind the sanctified authority of religion. When both entrench behind a false gravity, nothing remains but to subject them to the test of ridicule and to expose them." E. Cassirer, *The Platonic Renaissance in England*, translated by James P. Pettegrove, Nelson, 1953, pp. 183–184 (ECW 14, pp. 353–354).

[16]Hermann Cohen, in his *Æsthetics of pure Feeling* of 1912, already emphasized the relation humor as a form of pure sentiment and an esthetic apprehension of nature. He noted, in particular, based on the traditional opposition between beauty and sublimity, that nature was only beautiful by the intervention of humor: "[...] Nature in itself and by itself is not at all beautiful: only pure feeling introduces beauty in it; and only humor wakes up and enhances the beauty in nature." (*Ästhetik des reinen Gefühls*, Band 1, B. Cassirer, Berlin, 1912, p. 343).

[17]E. Cassirer, "Eidos and Eidolon: The Problem of Beauty and Art in the Dialogues of Plato" in E. Cassirer, *The Warburg Years (1919–1933)*, p. 217 (ECW 16, p. 138): "Galileo and Kepler are steeped in the same temperament of thought that runs through Plato's late dialogues and in particular, that found its expression in the *Timaeus* and in the *Philebus*."

[18]E. Cassirer, "Eidos and Eidolon: The Problem of Beauty and Art in the Dialogues of Plato" in E. Cassirer, *The Wrburg Years (1919–1933)*, p. 217 (ECW 16, pp. 137–138): "For no philosophical theory has ever more vigorously and more fully begun from aesthetic *effects* than this system, which abnegates a separate, independent and valid *existence* to aesthetics. [...]. If Platonism, however, had expressed the same force in the history of *science*, if, in particular, the founders of modern mathematical physics had declared themselves to be his students, they would only have taken up some motives that are already described more clearly in Plato himself."

[19]E. Cassirer, *The Individual and the Cosmos in Renaissance Philosophy*, Philadelphia, University of Pennsylvania Press, 1972, p. 163 (ECW 14, p. 188): "The scientific theory of *experience*, in the version to be given by Galileo and Kepler, will base itself on the basic concept and on the basic requirement of 'exactness' as formulated and established by the theory of art. And both the theory of art and the theory of exact scientific knowledge run through exactly the same phases of thought." He continues pp. 164–165 (ECW 14, pp. 190–191): "And it was only through this path that the Renaissance succeeded in overcoming magic and mysticism as well as the whole complex of 'occult' sciences. The alliance of mathematics and art theory accomplished what had

of nature to the forefront, as opposed to the notions of grammatical category and of rule of usage.

It is thus on the basis of a critique internal to the language sciences—grammar, rhetoric and logic which at the time constituted the "spoken arts", or the scholastic *trivium*—that the ulterior emergence of mathematical physics, as a mathematical theorization of nature, is to be understood.

1.1.1 The Critique of the Trivium: From Substance to Signs

Let us first begin by briefly recalling three points concerning the Aristotelian theory of knowledge, at least as it had been interpreted by late medieval scholastics and in the manner in which Cassirer related the debates surrounding it: the role played by *sensation* as a starting point for knowledge, the place occupied by *abstraction* making generality possible, and the primacy conferred to *finality* in the study of empirical reality.

The Aristotelian theory of knowledge had attempted to reproduce the structure of being—its forms and that to which it tends—using general ideas. In doing so, it had been led to distinguish matter and form in things perceived, the form being an object of knowledge, and matter representing a necessary condition for the existence of the thing.[20] In late scholasticism, the question arose regarding the participation of these two elements, sensation and intellect, in the constitution of knowledge. Very intense debates during the whole scholastic period had highlighted a certain number of shortcomings in the Aristotelian theory concerning the participation of these two elements; during the Renaissance emerged a coherent critique of the theory of *abstraction* which failed to account for the methodological function of the *concept* in a functional sense.[21] It is therefore a deeper examination of the question of the nature of conceptual knowledge that supports a certain number of critical points regarding medieval scholastics.

From the standpoint of the various sciences dealing with language, the critique put forth concerning the Aristotelianism of scholastics had fuelled debates regarding the distinction in principle to make between *words*, *things* and *faculties*. Employing the terms of a problem which had arisen from the very onset of scholasticism as

---

been impossible to achieve through dedication to empirical and sensible observation and through direct immersion of the feelings into the 'inners of nature'."

[20]E. Cassirer, *Das Erkenntnisproblem…*, Band 1, ECW 2, p. 120: "Thus the interpretation of the knowledge process still depends upon the realist presupposition at the foundation of the system: general concepts that are the last and highest achievements have a value because they find their counterparts in the general forms and purposes which organize and control empirical reality."

[21]E. Cassirer, *Das Erkenntnisproblem…*, Band 1, ECW 2, p. 125; as noted by Cassirer, Leibniz had re-edited in 1671 the work of the logician Mario Nizzoli [1498–1566] published in 1555 and which had already severely criticized the substantialism of the scholastic theory of knowledge without however managing to pave the way for a truly conceptual theory for it, failing to have sought in reason itself and not in an inductive generality the truly universal principles upon which to base knowledge.

part of the "problem of universals"[22] and which had for object the relation between words and things, Renaissance humanists sought to remove realism from the terms of the debate. By seeking to corroborate the *conceptual* character of knowledge, they criticized Aristotelian scholasticism for its tendency towards a hypostasis of the qualities conferred to objects but especially to faculties by attributing them a status as fully real, thus introducing a generalized confusion regarding the ontological status of the notions being handled. For the Renaissance humanists and contrarily to medieval thinkers, the faculties specific to thought are not substances which should have a same ontological status as the reality they allow to apprehend.[23] By breaking away from an epistemological attitude which does not have the means of avoiding the uncontrolled transformation of the logical method into ontology, it is then a matter for them of finding a *conceptual* means of specifying the role of faculties in the elaboration of knowledge without making them real entities. For the Renaissance humanists, this quest did not directly involve mathematics in the establishment of a functional solidarity between conceptual means and objects to be known—as will be the case later on during the seventeenth century—but firstly involved *textual criticism*, based on a new apprehension of the nature of language.

It is no longer a matter of questioning the reality of meanings attached to words but rather of finding within signs, thanks to the support provided by writing, a *material stratum subject to history* which can be the object of philological knowledge and which, by setting word usage in a historical context and in particular practices, accounts for their meaning. The difficulty of such an undertaking with respect to knowledge regarding texts is apparent: that words, the very instruments of signification, can be viewed as simple *signs* and not as the means of enriching an ontology with artificial entities, already indicates a truly functional approach to the nature of language.[24]

From this, two consequences follow and they concern the notions of style and of pedagogy inherited from scholasticism. Firstly, humanists wish to return to classical Latin and its Ciceronian model in place of the "degraded" Latin of the medieval era and the scholastic categories it conveyed. The constitution of a norm specific to "classicism" may then appear as being common to Latin and vernacular languages,

---

[22]By convention, the origin of this quarrel is set at 1108 with the publication of a text by Peter Abelard [1079–1142].

[23]E. Cassirer, *Das Erkenntnisproblem...*, Band 1, ECW 2, p. 101: "The predominance of abstract substantives is typical of a conception of nature and of spirit which transforms all properties and operations into real substances." This obviously reminds the critique addressed by William of Occam [1285–1347] to this type of epistemological attitude. Rarely cited by Cassirer, his critique of the realism conferred to ideas remains, according to Cassirer, in the orbit of scholasticism because it is not coupled with a theory of the *reality* of processes specific to reason. Cf. E. Cassirer, *Das Erkenntnisproblem...*, Band 1, ECW 2, pp. 210–211: "If nominalism rejects general concepts as fictions, as 'entia rationis' [beings of reason], this judgement as it is now expressed is not about Reason itself whose operations and capacities are absolutely considered as *realities*."

[24]As noted by Cassirer in *Substance and Function* p. 9 (ECW 6, pp. 7–8): "The reality of "universals" was in question. But what was beyond all doubt, as if by tacit agreement of the conflicting parties, was just this: that the concept was to be conceived as a universal genus, as the common element in a series of similar or resembling particular things."

as opposed to medieval Latin and its scholastic categories. It then becomes possible to oppose styles (classicism and scholastics) rather than contents (Latin and vernacular) and to thereby submit Latin to a philological reading: Latin (but also Greek, the language of the Gospels) is, *as is any other language*, submitted to historical evolution, which makes room for the notion of *historical textual criticism*. Thus, Lorenzo Valla was able to start questioning, on the basis of purely philological criteria, the authenticity of *sacred or secular* texts, by submitting *all of them without distinction* to the same critical rules.[25] Only textual *reality* thenceforth served as authority and the principles which guided its analysis needed to engage purely philological knowledge without it being submitted to a supreme science declared to be universal to begin with, a role attributed to logic in the *Trivium* curriculum.[26] Secondly, the pedagogy of the *Trivium* is profoundly modified by this, inasmuch as philological knowledge solicits the linguistic practice of the *vernacular languages*, which cannot be circumvented and which precedes any science of language proper. Also, the privilege conferred to Latin rhetoric began to be questioned: it is the objects of nature (of which natural languages form part) and the means of accounting for them which thenceforth became the focus of attention.[27]

1.1.2   The Status of Law

The second critique advanced against scholastic Aristotelianism concerns the theory of abstraction and the status conferred to universality stemming from it. In the eyes of humanists, the generality attributed to a proposition is marred with a major defect: since it is impossible to review all particular cases, abstract generality is never general enough to not be contradicted using the first counter-example that presents itself. It is thus necessary to conceive otherwise the relation between the particular and the general, by making use of the notion of *law*: for example, the diversity of sensible

---

[25]E. Cassirer, *Das Erkenntnisproblem…*, Band 1, ECW 2, pp. 101–102, concerning Lorenzo Valla: "For him, philology is not a goal per se, it is not a culture for the lettered, closed upon itself and self-sufficient, rather it always means for him the fundamental tool to discover the living, spiritual reality. […]. Should he discover that there are mistakes in the Vulgate or contradictions in the historical tradition of Livy's *Roman History* or should he track down the origin of the Donation of Constantine or of the Nicene Creed, it is always not so much the thing in itself than the joy of the critical and liberating activity that he enjoys and that serves as an incentive."

[26]E. Cassirer, *Das Erkenntnisproblem…*, Band 1, ECW 2, pp. 105–106 concerning Lorenzo Vives' critique of logic as a universal science: "The task […] expected from logic can only be fulfilled and solved by the totality of the particular sciences. Sciences must give themselves their own foundations; […]. The apparently general and formal content that dialectics develops is really based all over on hidden essential presuppositions that it extracts from the real sciences."

[27]For example, concerning Lodovico Vives who was the first to question the medieval pedagogy which was based on the purity of language rather than on the study of natural realities, Cassirer comments in E. Cassirer, *Das Erkenntnisproblem…*, Band 1, ECW 2, p. 104: "The dialectics, as a science of signs, should not degenerate into self-preoccupation; it can only win back its value and its relative right when it is satisfied with being used as the means and preparation for the knowledge of the object."

experience does not truly become *nature* unless it is put into correspondence with a unity founded by a law[28] rather than with categories which assorts this diversity into genera and species on the basis of their resemblance, as was the case since Aristotle[29] and his medieval following. Hence, what may suitably be called "nature" is no longer the kinship between logic and the classification of organisms into genera and species.[30]

Henceforth, knowledge does not aim so much to hierarchically classify things than to determine by means of laws their *reciprocal coherence*: even if this coherence does not yet express itself in a fully functional and mathematized way,[31] what makes objective determination possible is indeed already the idea that nature forms a *regulated system* and that it is by reaching this systematic viewpoint that it becomes possible to access an immutable nature based on which particular phenomena become determinable.[32] The idea of nature then carries a new value, that of the access to universality.[33] This access does not start by soliciting mathematical means: the systematic character of the knowledge of nature continues to be most often expressed using the notion of *soul of the world*. It is however necessary to interpret this notion

[28]Such is the case with Giordano Bruno [1548–1600]; E. Cassirer, *Das Erkenntnisproblem...*, Band 1, ECW 2, p. 244: "Therefore the idea of the circle supposes first to conceive an isolated point, then to run through an unlimited multiplicity of points while thinking at the same time at the law which controls the relationship between one another's particular positions and the common center."

[29]E. Cassirer, *Das Erkenntnisproblem...*, Band 1, ECW 2, p. 372: "The ideal of knowledge to which this conception corresponds is in fact the systematic classification of objects: it consists in differentiating various natural "forms" from one another and classifying their properties in a precise order."

[30]E. Cassirer, *The Problem of Knowledge...*; vol. 4, p. 124 (ECW 5, p. 144): "From the time of Aristotle there had been one point of contact and even more, an inner bond between logic and biology. Aristotelian logic was a logic of class concepts, and such class concepts were an indispensable means to knowledge and scientific description of natural forms."

[31]Concerning Campanella [1568–1639] and Galileo, Cassirer notes in E. Cassirer, *Das Erkenntnisproblem...*, Band 1, ECW 2, pp. 205–206: "With him [Galieo], he [Campanella] likes to repeat that Philosophy is only written in the book of Nature which lies open before our eyes; but the characters with the help of which we decipher it are not for him the "lines, triangles and circles" but rather the subjective qualities and perceptions of the different senses." Likewise, concerning Zabarella [1532–1589], Cassirer remarks in E. Cassirer, *Das Erkenntnisproblem...*, Band 1, ECW 2, pp. 115–116: "Whereas the mere empirical observation, in order to reach any conclusion at all, should require to get through all cases, what is specific in the process of science is that, through it, our mind immediately discovers and dicerns, in particular examples, the general law of its fundamental coherence: a law which the mind then applies back to particular facts and which reinforces the law through them. All these results come from Galileo's methodology in which we find again similar developments, except on one specific point which makes definitely the whole thing different. The role played by mathematics in the "demonstrative induction" is nowhere understood by Zabarella: the examples on which he dwells to establish his new fundamental conception are not selected from exact science which was only at hand in scattered statements but from the Metaphysics and Physics of Aristotle."

[32]E. Cassirer, *Das Erkenntnisproblem...*, Band 1, ECW 2, p. 249: "What is cognizable in a strict sense is never therefore the *thing of Nature* but it is *Nature* as the unique foundation and universal rule on which all phenomena are based."

[33]Whereas, during Antiquity, access to universality (under the guise of the divine) was mediated by the social structures and foremost by the *cities* then by the city of the universal God once the

as expressing for the first time the idea according to which nature forms a *whole sub-mitted to a legality which is not extrinsic to it*.[34] Indeed, as curious as it may appear to us who are used to a conflict between mechanism and finality, the supposition of a soul of the world aims on the contrary, during the Renaissance, to *found the idea of a mechanism of nature* which is independent from transcendental causes[35]: nature does not tend, by an ever-increasing hierarchy of its forms, towards an eminent exteriority which would play the role of ultimate cause but expresses *in itself* a universal harmony of which knowledge is capable of revealing the reasons.

1.1.3 The Role Conferred to Geometry

The third point concerns the place attributed to mathematics in their relation to the language sciences, because one may indeed ponder why the adoption of a mathematized language would by itself have the power to reform the *Trivium* from within and to derive conceptual knowledge from it. It is here that one of the possibly most curious aspects of modern thought intervenes: the scholastic epistemological attitude which limited itself, in the eyes of the humanists, to an uncontrolled relationship between logic and ontology, is substituted with a new attitude allowing for a third term enabling to escape such a sterile alternative. This third term introducing mediation between logic and ontology, such is precisely the status conferred to *geometry*. The latter begins to be considered as the sole purveyor of the precise determination of the *relations* between objects, precisely because it does not limit itself to these objects. Thus, for Campanella [1568–1639], if there indeed exists a split between sensible objects and their geometrical figuration, it is not due to an inherent defect of mathematics but rather to the fact that consideration was not given to that to which geometrical propositions truly refer, that is, *not sensible objects,* but rather *pure space*, existing as an immobile and incorporeal receptacle for all material things.[36] Conferring a *reality* to this pure space means interpreting geometry not as the copy

---

Roman Empire was formed, the Renaissance, for its part, witnessed the appearance of the notion that cults and articles of faith, owing to their always particular character, only provide a skewed image of God. Cf. E. Cassirer, *Das Erkenntnisproblem…*, Band 1, ECW 2, p. 223: "The unicity and permanence of nature warrant the unity of the true idea of God which neither specific cult nor specific dogma will exhaust or reach." It is therefore no longer through the city that the divine may reveal itself to mankind but through *nature* conceived as universal, precisely because it is devoid of any relationship with the exceedingly human character of societies.

[34] E. Cassirer, *Das Erkenntnisproblem…*, Band 1, ECW 2, p. 173: "The concept of the world-organism which is reached here is the first form the thought of a self-sufficiency of natural law takes on. From now on, no change—be it human or "demonic"—can be made through foreign arbitrariness unless it is not determined and prescribed at the same time through *proper* conditions which lie in the present state of things and their inner law of development."

[35] E. Cassirer, *Das Erkenntnisproblem…*, Band 1, ECW 2, p. 173: "Mechanism itself and its all-embracing validity cannot therefore be thought of without projecting a soul onto the cosmos."

[36] E. Cassirer, *Das Erkenntnisproblem…*, Band 1, ECW 2, p. 213: "The whole truth of pure mathematical configurations rests upon the fact that real counterparts stand by them in the world of pure space."

of a pre-existing reality but as that by which what is real in what is sensible becomes accessible to knowledge[37], thereby reconnecting with all which made the strength of the Platonic theory of Ideas.[38]

The Renaissance period[39] thus re-centers the question of the nature of space around the question of the value of science: space fits into *no category* according to which sensible objects are characterized—substance, quantity, or other—but it can nevertheless receive *any form* all the while itself being *uniform* since any part of space always behaves in the same manner, contrarily to the way in which Aristotle interpreted it by distinguishing qualitatively distinct natural "places". This Aristotelian theory of place made inconceivable any usage of the notion of *force* to account for the differences in object forms and movements since any object naturally tended to find its place.[40] Also, the theory of space, that is, geometry, must now be considered a *precondition* for any theory of physics because the notion of space is no longer abstracted from sensible objects but to the contrary makes their determination possible.[41]

1.1.4 Geometrical Physics

Thus, it is especially in the usage of geometry in physics that the greatest differences with Antiquity and the Middle Ages are felt. It is important to note that this new role conferred to geometry in physics fully coincides with the renewal of the language sciences. It is also a matter of being able to distinguish, under the stratum of mathematical *signification* as such, a *physical* stratum constituting the sole object of physical knowledge, as in the language sciences where it was necessary to distinguish the stratum of signification from the physical and historical stratum of signs.

---

[37] E. Cassirer, *Das Erkenntnisproblem...*, Band 1, ECW 2, p. 312: "Since the aspect of variation is here taken into account while universal and necessary validity is nevertheless maintained, the "Idea" is for the first time revealed and made productive in the range of empirical and concrete reality."

[38] Such was the case with Nicholas of Cusa and was still so with Marsile Ficin [1433–1499] for whom mathematics return to the place they had for Plato; E. Cassirer, *Das Erkenntnisproblem...*, Band 1, ECW 2, p. 76: "In [the prerequisites and foundations] we hold the ideal rules—as is clearly shown by the example of mathematics—; through them, we put to test perception and its exactitude and, consequently, they cannot have sensations and sensible objects as their limits and measurement."

[39] In particular, starting with Bernardino Telesio [1509–1588] and Francesco Patrizi [1529–1597].

[40] E. Cassirer, *Das Erkenntnisproblem...*, Band 1 (ECW 2, p. 215: "Anywhere where we see bodies trying to get closer to one another and striving to get in mutual contact, we have to attribute this only to specific determined forces that are active in them, not to a general tendency which is oriented towards avoiding or removing the empty space."

[41] E. Cassirer, *Das Erkenntnisproblem...*, Band 1, ECW 2, p. 219: "Just as space precedes matter, so is the knowledge of space more primary and certain than the knowledge of material bodies. The concept of extension that we put at the base of Geometry is not borrowed from material isolated objects through abstraction but it is on the contrary the condition under which we can only set up and observe the specific finite objects."

The new usage of mathematics in physics firstly entails a radical critique of its usage as being only *descriptive* of reality. As remarked by Kepler [1571–1630], any usage of geometry which aims solely to perform a *description* of physical phenomena is but outer garb which ought to be discarded because it does not involve in itself causal knowledge and hence thereby resembles a simple *allegory*.[42] Referring to work in the field of the history of science regarding Kepler,[43] Cassirer mentions the example of the ancient geocentric system, famous in the history of astronomy: such a system involved a description of an extreme geometrical sophistication using the theory of epicycles (a point revolving around the circumference of a circle having the Earth at its center) in order to justify the perception of retrograde planetary motions, which was enigmatic as seen from the Earth. The mathematical sophistication of the description of epicycles was such that astronomers, from the Hellenistic period onwards, had renounced conferring it any physical meaning.[44] The geometrical model then only appeared as nothing but the outer dressing of phenomena while saying nothing about their effective and causal functioning. The adoption of the heliocentric model by Copernicus [1473–1543] would expunge these inoperative complex geometrical descriptions which had been formulated in order to respond to teleological considerations concerning the "perfection" or "imperfection" of such or such "place" (the "superlunary" as opposed to the "sublunary") or of such or such "motion" (circular motion as opposed to any other type of motion). The usage of these "final causes" introduced normative considerations in physics and was thus, to the contrary, banished from a modern point of view. In the end, it thus appears, for Cassirer, that the *usage of mathematics in physics is not it itself a guarantee of scientificity.* What guarantees truly physical scientificity is *attributing to geometry the causal link itself* such as described by a *law* holding for nature *in general* and of which the application is not restricted to a region of experience, be it astronomical or

---

[42]E. Cassirer, *Das Erkenntnisproblem...*, Band 1 (ECW 2, p. 290: "The attempt to reduce reality to pure numerical relationships leads to mere allegorical games if, right from the start, it does not serve the rigorous causal analysis of natural processes, if it does not teach to see mathematics as a condition for empirical lawful knowledge and to use it accordingly. Kepler himself had perceived and expressed the historical duty which at this point fell to him with unsurpassable clarity."

[43]In particular, the works of E. F. Apelt [1812–1859] concerning Kepler, but also those of Pierre Duhem [1861–1916] regarding physical theory as well as ancient and medieval astronomy. The book by Alexandre Koyré *From the Closed World to the Infinite Universe*, which follows the same chronology as that of Cassirer for the modern period (from Nicholas of Cusa to Newton) and which marks a turning point in the study of this period, was only published in 1957, that is, twelve years after the death of Cassirer.

[44]E. Cassirer, *Das Erkenntnisproblem...*, Band 1, ECW 2, p. 285: "The number of the fixed spheres which would telescope one another would accumulate more and more—25 are mentioned by Eudoxus and Calippus and Aristotle raised the number to 49—without contributing to any exact description of the phenomena. That is why the astronomers from Alexandria had given up on any closer physical clarification of celestial phenomena and would do with the lessons of the epicycle theory [...]. The "truth" of this theory was never questioned then nor its capacity to determine the elementary components in which the theory analyses the movements of the stars in their empirical efficacy and efficiency."

otherwise.[45] The solely *descriptive* usage of geometry is thus completely *symbolic*,[46] in the pejorative sense where it only produces a sensible picture—even if such a picture is mathematically complex such as in the case of the theory of epicycles—of a reality of which the mode of apprehension, as geometrical as it may be, makes impossible the determined access to its causal nature. We clearly see that a solely symbolic usage of geometry contrasts with a thoroughly conceptual usage, inasmuch as such symbolic usage tends to radically separate the model from what it describes (as it did with Alexandrian astronomers), that is, to attribute a truly ontological content to mathematics (as did the Pythagoreans) whereas it is a matter of going beyond this alternative: it is now a matter of going beyond the meaning attached to the isolated perception of objects and to account for the *connection* between phenomena by the mediation of an abstract geometrical space. Within Cassirer, the *functional* approach in the natural sciences as opposed to a *substantial* approach finds itself defined along three points.

Firstly, it is not only a matter of developing a "functional" approach to reality that would simply seek to produce an exterior description of it; it is necessary to achieve a description in which the notion of ordered correspondence between elements enables to integrally explicate the *causal* schema of the contemplated phenomenon. Kepler notes this with respect to astronomy: it is from the moment when he realized that the velocity of planetary motion was mathematically proportional to the distance from the Sun[47] that he understood the nature of the *causality* linking the two phenomena, expressible in terms of determined quantities, even if the phenomena's physical nature as such still eluded him. By "physical" nature of the relation between phenomena, one must not understand such a relation to be "corporeal" or "material" itself, but only that it establishes a *norm regulating the totality* of the corporeal or material system in question (the Sun and the planets representing only a particular case), expressible in mathematical terms. From this point of view, even though, for us, the notion of cause implies the use of mathematical and physical concepts of which Kepler just began to reckon the scope,[48] it is indeed a causal schema which tends

---

[45] E. Cassirer, *Das Erkenntnisproblem...*, Band 1, ECW 2, p. 286: "In order for a hypothesis to be "true", it is not enough that it solely expresses in a short formula the astronomical phenomena which constitute only a limited portion of our total experience: it must at the same time render them in such a way as to actually correspond to our understanding of the conditions of all concrete natural events. The foundation of astronomy can only be achieved in connection with the scientific foundation of physics."

[46] As Kepler said in a letter to Joachim Tanck dated May 12th, 1608, cited in E. Cassirer, *Das Erkenntnisproblem...*, Band 1, ECW 2, p. 290: "For through symbols, nothing is demonstrated; no secret of Nature is revealed through geometrical symbols nor brought to light. They only give us results that were already known; unless is substantiated through secure grounds the fact that they are not merely allegories but that they express the way and the cause of the *connection* between two things that are compared."

[47] Kepler cited in E. Cassirer, *Das Erkenntnisproblem...*, Band 1, ECW 2, p. 295: "When I considered that the cause of the movement of planets would decrease with the distance from the sun the same way as light becomes weaker, I then concluded that this cause must have been something corporal."

[48] This is what differential and integral calculus will become from a mathematical standpoint and what the concepts of mass and of force will become from a physical standpoint.

to express itself in his works because the role attributed to functional relations is epistemologically considered to be *effective*. Thus, causality henceforth appears as a possible interpretation of the functional schema which was being epistemologically established at the time[49]:

> Thus side by side with series of similars in whose individual members a common element uniformly recurs, we may place series in which between each member and the succeeding member there prevails a certain degree of difference. Thus we can conceive members of series ordered according to equality or inequality, number and magnitude, spatial and temporal relations, or causal dependence.

Secondly, the functional approach in the natural sciences which begins to develop with Kepler confers primacy to *geometry* over any other mathematical field, including arithmetics. The reason for this is clear: numbers as such do not have *any* particular power for describing the causal schema; on the other hand, it is indeed the geometrical form which, by distinguishing the elements present and by tracing a framework within which their relationships are set, enables to confer a *physical* sense to quantities all the while avoiding these being nothing more than superficial veneers of entities taken in isolation. This primacy of geometry in physics does not come from the figural character of geometry because it is precisely by abandoning any attempt at figuration that Kepler has managed to question the idea that planetary orbits had to be circular. What remains constant in the trajectory of the orbit is precisely not its *figure*— which, as an ellipse, appears as variable—but rather the action of the reciprocal *force* of attraction between the Sun and the planet in question.[50] It is thus a new division between permanence and variation which is in effect in the modern usage of geometry: far from the ancient attitude which consisted in considering as belonging to concepts only that which is permanent as opposed to empirical variation, modern geometrical attitudes consist on the contrary in apprehending variation on the basis of principles of *variability* which, alone, are permanent. It is this new articulation which makes possible the truly conceptual description of material phenomena.

Thirdly, it is clearer where the relation between the sciences concerned with language and the natural sciences lies from a modern standpoint: in both cases, it is a matter of uncovering, under the stratum of the meaning of words or of perceptions, *rules* accounting for the *connection* between phenomena according to their specific mode of presentation, founded, in the case of the language sciences, upon the *repetition* of usage forms and, in the case of the natural sciences, upon the geometrical *permanence* of the correspondence between elements in the causal schema. In the remainder of these works, Cassirer will forget neither this relatedness nor the chronological anteriority he confers to the language sciences of the Renaissance in the

---

[49]E. Cassirer, *Substance and Function*, p. 16 (ECW 6, p. 15).

[50]E. Cassirer, *Das Erkenntnisproblem…*, Band 1, ECW 2, p. 310: "For Kepler, […] the constant which is looked for does not lie in the form of the trajectory anymore but rather in the *principles* of his mechanics and physics: what is constant is the action of the force of attraction as well as the "magnetic" power of the sun, even though another numerical value is attributed to it at each point of the trajectory. The unambiguous functional law which binds up the quintessence of infinitely possible variations as "united in a group" and determines the path of planets with more precision than the fictitious celestal cercles have ever been able to."

scientific revolution involved in producing the modern viewpoint. Such a viewpoint develops through two other stages, one occurring during the seventeenth century and habitually designated as "classical science", and the other one corresponding to the eighteenth century and designated as "*Mathesis Universalis*".

1.2 Classical Science

With classical science having, according to Cassirer, for most eminent representatives Galileo [1564–1642] and Descartes [1596–1650], begins a second stage of modern science following the Renaissance. Cassirer essentially describes the transformations having taken place during that period touching the natural sciences and he only addresses those affecting the language sciences when referring to the end of the period (with Descartes) and does this in order to introduce the change in perspective initiated by Leibniz.

It is Galileo who opened up a new theoretical field[51] within which are studied the *quantitative relationships between variables* without presuming, as did Kepler, the *meaning* that these quantitative relationships could have in terms of relations between empirical objects, which would have involved returning to the idea of a harmony and of a "soul of the world". The point of view of classical science (Galileo and Descartes) therefore consists in dispensing with any *teleology* and in constituting the idea of knowledge without recourse to the idea of perfection: the fields of knowledge and of values no longer bear a direct relation. The new point of view thus develops a functional analysis of which the specificity is to question all notions considered to be intuitively and empirically opposed, notions such as finite/infinite, straight line/curve, motion/rest. It is the idea of the *relativity* of points of view which is then put forth, any absolute notion concerning space, time, numbers or motion being progressively abandoned to the benefit of formal relationships of which mathematics *demonstrate* the reciprocal relations and, thereby, the necessary character. The questioning of the aforementioned differences deemed to be absolute thus contributes in transforming the very idea of science: the latter would then consist in the movement aiming to surmount the rigid oppositions, be it in the concepts employed or in the relation between the two disciplines which are mathematics and physics. This movement rests on a profound renewal of the very nature of what is to be understood by "mathematics": as shown by Descartes, it no longer has the status of a singular science among many but must rather be viewed as a new means for the *general* apprehension of knowledge,

---

[51] Cassirer notes however the capital legacy for Leonardo da Vinci regarding the elaboration of the Galilean point of view; E. Cassirer, *The Individual and the Cosmos in Renaissance Philosophy*, Philadelphia, University of Pennsylvania Press, 1972, p. 156 (ECW 14, p. 180): "Leonardo's formulation of *individual* laws of nature may sometimes be vague and ambiguous; but he is always certain about the idea and the definition of the law of nature itself. On this point, Galileo bases himself directly on Leonardo, simply continuing and explicating what the latter had begun. For Galileo too, nature does not so much 'have' necessity, but rather *is* necessity."

exclusively founded on the notions of order and measurement.[52] This is what enables to constitute a universal research method which does not depend on the particular objects under study but which rather fully rests upon the unity of the intellect.[53]

In this new context, it is important to note a change in the usage of an adjective which will be later developed to a great extent by Cassirer, that of "symbolic". Having until then been connected to the critique formulated by Kepler concerning the allegorical character of the usage of numbers and geometrical figures,[54] or by Descartes concerning the muddled and methodless[55] character of the ancient and medieval sciences, and conceived by Cassirer in an even more general manner as referring to all practices within which the usage of signs aims to unveil a mystery,[56] the term "symbolic" acquires, within the context of the analysis of classical science, a whole other meaning: Cassirer indeed begins to describe the laws of motion uncovered by Galileo as *"symbolic* units"[57] enabling to link to a shared calculation the quantities representing empirically *non comparable* objects, such as space, time, and velocity. The adjective "symbolic" thus refers for him to the type of theoretical construction unveiled for the first time by Galileo[58] and which pertains to *space* as it was conceived according to seventeenth century science. The characteristic of this new type of symbolic construction is, according to Cassirer, that it defines its *own domain of*

---

[52]E. Cassirer, *Das Erkenntnisproblem…*, Band 1, ECW 2, p. 371: "One can try to determine this order in figures or in numbers, in the stars or in the sounds; always remains as the unique starting point the universal idea of connection and correlation. A pure science of *proportions* and *relations*— independent from any specificity of the *objects* in which they present and embody themselves— constitutes therefore the first requirement and the first object on which the method adjusts itself."

[53]E. Cassirer, *Das Erkenntnisproblem…*, Band 1, ECW 2, p. 369: "We cannot know anything from the things, without at the same time becoming aware of the essence of our own thinking. The pure understanding constitutes the first object that is opposed to us in the series of truths."

[54]Cf. note 43 above which provides a citation of the letter addressed by Kepler to Joachim Tanck.

[55]E. Cassirer, *Das Erkenntnisproblem…*, Band 1, ECW 2, pp. 370–371: "[…] indeed because of their constant dependence on immediate sensible intuition and on their clumsy mode of symbolic notation, the geometrical analysis of the Ancient and the modern arithmetic offer a complicated technique which would confuse the mind rather than a fully transparent knowledge which clarifies and forms it."

[56]E. Cassirer, *The Philosophy of Symbolic Forms*, vol. 2, p. (ECW 12, p.): "We are accustomed to view these contents as "symbolic", to seek behind them another, hidden sense to which they mediately refer. Thus, myth becomes mystery: its true significance and depth lie not in what its configurations reveal but in what they conceal. […]. From this result the various types and trends of myth interpretation—the attempts to disclose the meaning, whether metaphysical or ethical, that is concealed in myth."

[57]E. Cassirer, *Das Erkenntnisproblem…*, Band 1, ECW 2, p. 362: "Thus, when Galileo somehow compares, in his deduction of parabolic throw, distances, times and impulses that he relates to a single line segment as a general symbolic unity for all these various magnitudes, he has applied the real method of "figurative" characteristic."

[58]As noted by M. Ferrari, this change of meaning attributed to the adjective "symbolic" appears as early as 1904 in the personal itinerary of Cassirer when he made his first recourse to the expression "symbolic form" while commenting Leibniz' *Monadology* (§ 61) in his edition of the works of Leibniz (G. W. Leibniz, *Hauptschriften zur Grundlegung der Philosophie*, II, p. 173 note 114). Cf. Massimo Ferrari, *Ernst Cassirer; dalla scuola di Marburgo alla filosofia della cultura*, Leo S. Olschki Editore, Firenze, 1996, pp. 174–175.

validity by elaborating new theoretical tools *which do not integrally depend upon those which preceded it.*[59] Cassirer thus develops, through an analysis which seeks to keep as close as possible with the effective transformations undergone by science, a new way of conceiving its *history* since it is no longer conceived according to a cumulative and linear model but as rather being made of *epistemological ruptures* which define theoretical frameworks having their own internal field of validity.

1.2.1 A New Variational Viewpoint in Mathematics

As pointed out by Cassirer, the usage of what may be called "synthetic" geometry, founded on Euclid's [325–265 BC] *Geometry* and on the *Conics* of Apollonius [262–190 BC], already began to prove inadequate during the age of Kepler. In the evolution of his astronomical descriptions, Kepler indeed progresses from a *static* viewpoint which remains in full accordance with the science of Antiquity (when he defines a geometrical constant such as the elliptical form of the orbit of a planet around the sun) to a *dynamic* viewpoint focusing on the *variation* of parameters at each instant (distance, velocity, area, revolution period, forming the content of Kepler's three laws of planetary motion[60]) undergone by continuous forces.[61] In later research, it is the geometrical forms themselves which he analyzes in terms of variation: the case of the circular orbit is thus interpreted as a limit case of the elliptical orbit. In the same line of thought, Galileo considers the curvilinear trajectory of projectiles as being composed of line segments[62]: the straight line and the curve are no longer opposed in an absolute manner because it is now possible to continuously transition from the one to the other.

Generally speaking, there emerges a need for a new calculation that is no longer limited to a proportional relation conceived of as fixed between variable quantities corresponding to empirical objects or places but which directly accounts, in the form of a generation law, for the infinite variability of an indeterminate number of quantities bearing relations of proportionality.[63] The notion of proportionality thus tends to evolve because the compared quantities are not fixed once and for all, but

---

[59] E. Cassirer, *Das Erkenntnisproblem...*, Band 1, ECW 2, p. 351: "The new goal that is taken into consideration generates the new means from within."

[60] That is, the existence of a plane for the planetary orbits as well as its elliptical nature, the area covered by the distance between the Sun and a planet at any moment of its orbital revolution or the period of such revolution.

[61] E. Cassirer, *Das Erkenntnisproblem...*, Band 1, ECW 2, p. 352: "Since this force depends on the distance of the planets and consequently varies from one point to another, in order for its intensity to be determined, means must be found to gather infinitely many different impulsions active at each moment in a unitary model, in a common size. Here the concept of the definite integral is uplifted to a clearer meaning [...]."

[62] E. Cassirer, *Das Erkenntnisproblem...*, Band 1, ECW 2, p. 349.

[63] E. Cassirer, *Substance and Function*, p. 17: "That which binds the elements of the series *a, b, c,...* together is not itself a new element, that was actually blended with them, but it is the rule of progression, which remains the same, no matter in which member it is represented."

rather continuously variable: the mathematical expression of the laws governing these continuous variations becomes the central focus.[64]

The capital issue is the shift that such an attitude involves with respect to the relation between fixed determination and variable determination in mathematics: variation becomes fully integrated under the perimeter of mathematical determination because its purpose is fully justified by principles of variability expressed through rigorously determined mathematical laws. Furthermore, continuous variations and the principle of variability become *reciprocal correlates*, as did the conceptual differences which had previously been deemed to be absolute, such as the one between straight and curved lines. These changes which make variations appear as conditions of determination and not as their limits go as far as to require a new conception of geometrical space.

It was Descartes who drew the truly epistemological lessons from the Galilean revolution, not so much by proposing novel geometrical problems than by introducing a new way to *represent* them; there is therefore indeed here what we could call a "meta-geometrical" endeavor having at its core the best way to represent objects using adequate *mathematical signs*. This new representation aims to manifest the unity of the geometrical method not using isolated problems with which he found fault in the geometry of the Ancients, but using problems which, despite appearing to be distant from one another, nevertheless present common solutions except for variations in parameters. The difficulty then consists in revealing a common measure—in the sense of establishing a relation—between parameters which, empirically, tend to be put into absolute opposition. On the contrary, from the standpoint of the new geometry advocated by Descartes, any concept is susceptible of *being put into a relation with its opposite*: for example, the opposition deemed to be absolute between discrete and continuous quantities can be overcome if, for a given type of problem, we manage to find the measurement unit thanks to which it becomes possible to resolve the continuous quantity into enumerable parts; likewise, a given problem's common measure can as easily be a straight line as it can be an indivisible element (a point), if we believe its repetition to be sufficient for constructing finite figures. Any isolatable parameter constitutes a "dimension" of the problem addressed, if by "dimension" we mean only the rule owing to which an object becomes measurable.[65] Also, beyond length, width, and height, the relevance of the concept of dimension extends to any parameter likely to be isolated, including in the physical domain within which weight (as a measurement of the gravity exerted) or of velocity (as a measurement of motion)

---

[64]E. Cassirer, *Das Erkenntnisproblem...*, Band 1, ECW 2, p. 353: "Here the fundamental and typical characteristic of mathematical thinking concerns the capacity to decompose each unity into an infinite multiplicity and to let multiplicity unify again into a totally determined conceptual image. Galileo's phrase that unity and infinity are correlated concepts should be understood this way."

[65]E. Cassirer, *Das Erkenntnisproblem...*, Band 1, ECW 2, p. 377: "The "dimension" designates the rule of thought in virtue of which an object is viewed as measurable. Hence, not only length, width and depth, but also weight fall into this category, as the matrix according to which the weight of bodies, the speed, and the measure of movement can be estimated and determined. Generally speaking, all these pieces of determination define an entity unambiguously and thereby contribute to making it distinguishable from all others, as "dimensions" of this entity.

can constitute the dimensions of a problem. Once such parameters have been clearly distinguished, interrelating them presupposes a theoretical domain within which they may be compared. For Descartes, as for Fermat, this theoretical domain always pertains to *space*, designated by the term "extension": any point can be determined therein using a system of coordinates formed by two arbitrarily defined straight lines and their associated numerical values.[66] It thereafter becomes possible using such a space to algebraically translate geometrical figures in a seamless way. Such a space is, according to Cassirer, considered by Descartes to form part of primitive ideas[67]: it is these ideas which enable to produce a judgment in function of order and measurement and which, thereby, precede the sensible by making its determination possible.

1.2.2 The Generativity Specific to Mathematical Physics

The modern viewpoint in physics breaks with the idea according to which a mathematical concept is to be "applied" to a preexisting reality and which would, for some unknown reason, be predisposed to receive its form: in fact, both terms of "concept" and "reality" *are themselves relative terms* and mathematical physics constitutes precisely the movement of thought seeking to bring this to light. The new role attributed to the mathematical concept thus does not consist in receiving a "content" that could be related according to the modality of resemblance but to detect in the deduction of the consequences of the mathematical concept a relationship with a schema of physical causality. The mathematical concept thus represents, to begin with, a simple hypothesis which does not concord at the onset with physical facts because it does not allow to perform measurements upon bodies. It is through mathematical series within which the use of numbers is rigorously determined that we manage to compare such series with those of the causality specific to bodies using the numerical form of a measurement. The measurement is therefore also the product of a serial relation which is precisely the only one to be comparable in the mathematical and

---

[66]E. Cassirer, *The Individual and the Cosmos in Renaissance Philosophy*, Philadelphia, University of Pennsylvania Press, 1972, p. 186 (ECW 14, p. 215): "Herein lies one of the most important advances of modern analytical geometry beyond Greek mathematics. In the latter, too, there are some very definite beginnings of the use of the concept of co-ordinates. But they are always concerned solely with an individual figure, given in each instance. The notion never raises itself to a truly general principle. [...]. It was Fermat who first created a method that was free of all such limitations and that permitted the centre of the relational system to be located anywhere in the plane of the curve. The direction of the abscissa and ordinate axes also allows all sorts of translations and rotations. Instead of perpendicular co-ordinates one can also use obliquely intersecting axes—in short, the system of co-ordinates is completely independent of the curve itself."

[67]E. Cassirer, *Das Erkenntnisproblem...*, Band 1, ECW 2, p. 384: "There are several primitive notions, each according to the various classes and problems that can become the object of our research: whereas some like Being, Number and Duration are valid for all contents, some others like Space, Form and Movement relate more specifically to bodies, while others, only to the soul."

physical fields.[68] Therefore, to put mathematics and physics into relation does not consist in applying a mathematical form to a physical content nor does it consist in seeking a hybrid mathematical/physical field but rather to radically separate the two orders of legality by posing the question of their correlation which takes the form of a function.

This manner of exploiting the generativity specific to the mathematical concept in view of physical knowledge has another consequence also: it implies no *a priori* limit to its field of relevance. Any field in which a schema of causality can be detected is therefore susceptible on an *a priori* basis to be mathematically determined, whether it forms part or not of the already constituted corpus of the natural sciences.[69] There lies an *a priori* principle of extension of the field of physics upon which philosophy, particularly Cartesian philosophy, will be required to ponder, inasmuch as it tends to favor a radically mechanistic conception of the world from which the domain of values first appears to be radically excluded.

1.2.3   The Relativity of Motion in Modern Physics

The relativity of mathematical concepts extends to those of physics and, foremost, to the notion of motion. Galileo, by considering physics as the study of the relations between quantifiable parameters, significantly deviated from the conception the ancient and medieval world had of it. This is particularly the case in what concerns the nature of motion: whereas, for Aristotle, motion is a *property* of bodies, it becomes a measurement with respect to a frame of reference for Galileo. Henceforth, uniform rectilinear motion becomes for Galileo the primary state of any body, immobility now being but a particular case of rectilinear motion.[70] If any body uniformly conserves its rectilinear motion without it needing the continuous application of a force, motion is thereby nevertheless the *result* of a *proportional composition*

---

[68] Such is the case, for example, with Galileo's concept of uniform acceleration; E. Cassirer, *Das Erkenntnisproblem...*, Band 1, ECW 2, p. 322 which quotes Galileo's *Dialogue Concerning the Two Chief World Systems*: "From now on, if the properties we have derived from the free fall of natural bodies are verified by experience, this allows us to conclude without danger of error that the concrete movement of fall is identical to the one we have defined and presupposed: should it not be the case, our proofs would indeed not lose anything of their force and of their conclusive power, given that they are only valid for our hypothesis—the fact that there is not one body in nature which is granted a spiral movement does not affect the propositions of Archimedes about the Spiral."

[69] E. Cassirer, *Das Erkenntnisproblem...*, Band 1, ECW 2, p. 322: "The goal of physics is precisely to keep in all its stringency power the concepts that mathematics provide and at the same time nevertheless make them productive for always larger domains of particular facts."

[70] E. Cassirer, *Einstein's Theory of Relativity*, p. 362 (ECW 10, p. 14): "Objective physical reality passes from place to change of place, to motion and the factors by which it is determined as a magnitude."

of forces (of impulse, friction, etc.), by virtue of an "arithmetic of forces".[71] For Galileo, it is in this sense that motion forms part of pure mathematics.[72]

The case of the calculation of projectile motion in ballistics is an example of this. Until then, it was not possible to conciliate within a body the possibility of two vertical and horizontal motions because of their differing directions: motion being inherent to the projected object, it was therefore not possible to make it into a combination of motions. But from the moment when projection is conceived as a decomposition of velocity into two components (vertical and horizontal) according to a common measure, that of the *uniform* nature of motion (uniformly accelerated vertical fall and uniform horizontal motion), it becomes possible to mathematically *deduce* the parabolical nature of the trajectory.[73]

Questioning the "natural" immobility of bodies at a physical scale also entails questioning the immobility of the Earth and its central position on an astronomical scale. Very progressively, over the course of the beginning of the seventeenth century, and owing to Galileo's mediation, the geocentric system of Ptolemy [90–168] would be substituted by that of Copernicus [1473–1543], which had been described one century before Galileo. This substitution tends to shift the attention from a notion of an empirical "fixed point" towards the notion of an *arbitrarily* set point in order to make astronomical motion intelligible: Copernicus and, following him, Kepler, began to *vary* the fixed point used to describe the motions of the planets, the sun, or fixed stars.[74] However, the full-fledged adoption of the notion of relativity of motion with respect to a reference point considered to be fixed only occured with Galileo from the moment when the velocity of a body was itself considered as what

---

[71]This is an expression used by Kepler, often cited by Cassirer, for example in E. Cassirer, *Das Erkenntnisproblem...*, Band 1, ECW 2, p. 300 where he speaks of the "arithmetic of fundamental forces" or in E. Cassirer, *Einstein's Theory of Relativity*, p. 362: "But in truth, no place in itself is opposed to any other, but there are in nature only differences in the mutual positions of bodies and of material masses. [...]. It is implied in this that what we call the "true place" is never given to us as an immediate sensuous property, but must be discovered on the basis of calculation and of the "arithmetic of forces" in the universe."

[72]E. Cassirer, *Substance and Function*, p. 175 (ECW 6, p. 197): "Galileo, at least, leaves no doubt that the principle, in the sense that he takes it, has not arisen from the consideration of a particular class of empirical real movements. [...]. The concept of uniform motion in a straight line is here introduced purely in an abstract phoronomic sense; it is not related to any material bodies, but merely to the ideal schemata offered by geometry and arithmetic. Whether the laws, which we deduce from such ideal conceptions, are applicable to the world of perception must be ultimately decided by experiment; the logical and mathematical meaning of the hypothetical laws is independent of this form of verification in the actually given".

[73]E. Cassirer, *Das Erkenntnisproblem...*, Band 1, ECW 2, pp. 348–349.

[74]E. Cassirer, *Das Erkenntnisproblem...*, Band 1 (ECW 2, p.): première edition, pas dans les Gesammelte Werke Le problème de la connaissance dans la philosophie et la science des temps modernes, tome 1, De Nicolas de Cues à Bayle, (2004) [1906 and 1910–1911], p. 476.

constituted the manifestation of its invariant aspect,[75] this being what enabled to define in a purely functional manner the notion of the relativity of motion.[76]

What, then, is understood by "geometry" in the context of classical science and what place does it occupy within the natural sciences? Cassirer notes that the citation by Galileo which is so often used and according to which the book of nature is written in mathematical language, using different characters than those of our alphabet because it is composed of "triangles, circles, and other geometrical figures",[77] does not adequately account for the epistemological situation of the modern period because, if it indeed focuses on the primacy of geometrical forms in the study of nature—as Kepler already did when seeking to describe the cosmos using the five regular polyhedra described by Plato in the *Timaeus*—it does not, on the other hand, account sufficiently for the transformation concerning the *meaning of space*: now conceived as fully Euclidean due to its homogeneous and isotropic character, endowed with an arithmetic structure which enables an indefinitely more precise location of the points composing it and which is likely to describe variations on the basis of principles of invariance in space, it allows for a full reconfiguration of the very notion of geometry.[78] It is the work undertaken by the thinkers of the *Mathesis Universalis*, that is, according to Cassirer, Leibniz [1646–1716] and Newton [1643–1727] which would fully achieve this mutation of mathematical knowledge into functional knowledge.

---

[75] E. Cassirer, *Einstein's Theory of Relativity*, p. 363 (ECW 10, pp. 14–15): "The new measure, which is found in inertia and in the concept of uniform acceleration, involves also a new determination of reality. In contrast with mere place, which is infinitely ambiguous and differs according to the choice of the system of reference, the inertial movement appears to be a truly intrinsic property of bodies, which belongs to them "in themselves" and without reference to a definite system of comparison and measurement. The velocity of a material system is more than a mere factor for calculation; it not only really belongs to the system but defines its reality since it determines its vis viva, i.e., the measure of its dynamic effectiveness."

[76] E. Cassirer, *Das Erkenntnisproblem...*, Band 1, ECW 2, p. 333: "In contrast, for modern knowledge, it is visibly the idea of necessity which constitutes the meaning and the content of the concept of nature. Hence modern knowledge does not conclude from the quality and composition of an isolated form of movement to its persistence and duration but, on the contrary, it is through the universal law of conservation that speed becomes a reality in the sense of knowledge, a permanent and determined 'essence', which only appears in the immediate, naive conception as an arbitrary emerging and disappearing content."

[77] Galileo, *Il Saggiatore*, Chap. 1, § 6.

[78] E. Cassirer, *Das Erkenntnisproblem...*, Band 1, ECW 2, p. 351: "However, how characteristic these propositions are, they are not sufficient to describe the whole of Galileo's activity and the real progress that is meant by it. The language of Nature, as it is defined here, is that of the antique, synthetic geometry. Euclid's *Elements* and, in a larger scope, Appolonius' theory of conic sections, fully belong to its grammar. It was enough to understand the Keplerian laws of planetary motion: it was not sufficient for the laws of falling bodies and the fundamental dynamical viewpoint which was built upon them. The task of Galileo was not limited to approaching the meaning of phenomena by the use of ready-made conceptual tools; it was requested for the new vision of Nature to first discover a new language, to determine its character and to build its syntactic structure according to fixed rules."

1.3 Mathesis Universalis

As already stated above,[79] Cassirer considers the modern era to have begun with a revolution in the language sciences which progressively spread towards the natural sciences via a renewal of the Platonic notion of Idea, itself conveyed in the aesthetic notion of style. But whereas the language sciences had not captured the same level of attention from the thinkers of the Classical period for reasons which, as we will see, stem from their epistemology, such a reflection on the language sciences is brought back to the forefront by the "Mathesis Universalis" project, particularly with Leibniz.

1.3.1 A New Approach to the Language Sciences

Indeed, during the Classical period and particularly with Descartes, allusions to the role of language in the development of knowledge are rare.[80] Cassirer particularly notes[81] that the letter addressed by Descartes to Father Mersenne on November 20, 1629, in which Descartes responds to Mersenne concerning the possible constitution of a universal language, mainly focuses on a difficulty in principle which tends to ruin the whole project. Indeed, for such a language to truly enter into usage, it would need to be superior to vernacular languages in terms of precision and in terms of the sequencing of certain ideas it would enable. However, for such a language to exist, the sequence of ideas would need to have already been fully established, meaning that the decomposition of ideas into simple ideas would operate following the numerical model. The project is therefore deemed by Descartes to be unrealistic as long as he has not fully accomplished the decomposition of ideas into simple ideas... a project he will never achieve, his *Rules for the Direction of the Mind* remaining unfinished.

It is nevertheless by tackling this difficulty with respect to principles that Leibniz will carry on the project and bring it to evolve. During his youth, the latter indeed sought to compose an "alphabet of human thoughts": it was a matter of decomposing thought into primitive thoughts using the model of the decomposition of numbers into prime factors, in the manner imagined by Descartes. This alphabet of human thoughts would then serve as a privileged instrument for constituting a universal "Encyclopedia", that is, the ordered knowledge of all things, composed on the one hand by the repertoire formed by all simple ideas and, on the other hand, by the way to deduce complex ideas from them. Thus, the "alphabet of human thoughts" is conceived by Leibniz so as to make possible an almost arithmetic increase in

---

[79]Cf. §§ 11 and 113 of this chapter.

[80]E. Cassirer, *The Philosophy of Symbolic Forms*, vol. 1, pp. 127–128 (ECW 11, p.): "Descartes, who provided the universal philosophical foundation for the Renaissance ideal of knowledge, saw the theory of language in a new light. In his principal systematic works Descartes gives us no independent philosophical study of language [...]."

[81]E. Cassirer, *The Philosophy of Symbolic Forms*, vol. 1, p. 128 (ECW 11, p.): "The ideal of the unity of knowledge, the *sapientia humana* which always remains one and the same, regardless of how many objects it encompasses, is here extended to language. Too the demand for a *mathesis universalis* is added the demand for a *lingua universalis*."

knowledge. In this perspective, any notion present in language can be decomposed into simple elements and any proposition can be interpreted as a relation between a subject and what qualifies it—its predicates—which are put into relation following the model of numerical decomposition: predicates are then said to be "contained within the subject" in the same manner as a number "contains" its prime factors.

However, Cassirer notes, this strongly Cartesian project had gradually transformed and a mature Leibniz would eventually tend to depreciate the notion of original simplicity conferred to the elementary, as he would devalue the concept of number as model for all knowledge. The notion of *relation* indeed takes precedence over the concept of number which is thenceforth conceived by Leibniz as a relation among others, in fact, the simplest one possible between two elements. The essence of any relation therefore lies not with the concept of number, but with *order*, be it numerical or not. Then, for Leibniz, it is no longer a matter of returning to the origins of thought in an "Encyclopedia" which would be capable of arithmetic growth since thought is no longer supposed to be reducible to the numerical model. It is a matter of instituting a universal *Mathesis* by constructing, for each particular domain (geometric location, quantity, force), formal rules enabling the constitution and ordered comparison of series of elements which are not inherently simple but which are only so with respect to a duly specified generation rule.[82] Thus, the same functional structure,[83] founded upon the notions of formal deduction and of computation appears in the varied but related fields which are geometry, arithmetics, and physics.

1.3.2 Towards a "General Science"

As remarked by Cassirer, when examining Leibniz's intellectual progression step-by-step, it becomes clear that he first sought to conceive all relations on the basis of the numerical model which he later renounced from the moment whence he considered, on the contrary, numbers to represent only the simplest case of the *functional relation* in general, the quantitative aspect being, in the end, considered to be secondary. From then on, he sought to develop a logic of functional relations which would be specific to each type of object considered, following which he finally came to see in the calculation of functions (and in particular in the differential and integral calculus) a general logic enabling to define both the arithmetic principle of numbers, the

---

[82]E. Cassirer, *Das Erkenntnisproblem...*, Band 2, ECW 3, p. 117: "So far the attention had primarily been fixed upon the determination of the elements from which the inner composition would be generated, but now it is above all turned towards the forms of connectedness."

[83]E. Cassirer, *Das Erkenntnisproblem...*, Band 2, ECW 3, p. 117: "From a general point of view, it is from now on the *concept of function* which replaces the concept of number as the real foundation and content of mathematics. The whole system of universal science evolves this way a characteristic transformation."

geometric principle of magnitude, and the physical principle of force.[84] These are but simple samples of a "general science" which itself remains to be founded.

Regarding geometry, Leibniz criticizes the Cartesian point of view because it fails to directly describe the notion of geometric locus but only translates it into semi-geometrical, semi-algebraic terms. On the contrary, it is necessary to *directly* analyze the notion of geometric locus by endeavoring to describe the decisive elements specific to each figure in order to achieve authentically geometric knowledge.[85] It then becomes necessary to develop a logic which would be specifically geometric and which would be thinkable outside of any quantification because the notion of quantity is not specific to this domain. To do this, it is not the elements which need to be conceived in a clear and distinct manner as Descartes sought to do, but rather the *rules of construction* from which can be deduced the formal properties defining the nature of the elements under consideration (for example, the two necessary and sufficient points enabling to define the notion of straight line). This relational principle is truly general in scope because it allows to fully account for the diversity of figures: directly incomparable amongst themselves using intuition when they are not of the same species (for example straight lines and curves), they nevertheless become comparable if viewed from the level of the infinitesimal change in direction of each point, determinable with precision using a genetic rule.[86] It is in this manner that a "universal mathematics" becomes conceivable: such a science is not identical for all domains of knowledge but it is on the contrary likely to vary in its deduction methods when a specific rule can be exhibited on the basis of which a conceptual domain and the formal properties of its elements is circumscribed.[87]

Regarding the field relative to quantity, once again, the mutation of the point of view operated by Leibniz is considerable because it becomes possible to apply the

---

[84] E. Cassirer, *Das Erkenntnisproblem...*, Band 2, ECW 3, pp. 128–129: "From the "Geometry of the indivisible" by Cavalieri which shows us "so to say the rudiments or beginnings of the lines and the figures", Leibniz goes to the physical concept of the moment of speed. From here he moves towards analytical Geometry and the "inverse problem of the tangent", in which however he is never stuck to one particular task as such, but to the universal method of its resolution."

[85] E. Cassirer, *Das Erkenntnisproblem...*, Band 2, ECW 3, p. 120: "All the diversity which is found in the particular figures that are intuitively conceived must be completely derivable from the difference between their fundamental logical moments."

[86] E. Cassirer, *Das Erkenntnisproblem...*, Band 2, ECW 3, p. 140: "Here we see how the differential which is not at all identical and equivalent to a figure, can nevertheless represent its total conceptual meaning and bring to an exact expression all the relationships it has with other magnitudes."

[87] E. Cassirer, *Das Erkenntnisproblem...*, Band 2, ECW 3, pp. 120–121: "The Analysis of Situations fulfills the general task that Leibniz has set forth in his Universal Science: it breaks up the ready-made formations of the mind in a movement of thinking that develops according to rigorous rules and determines its final result from the formal features of this development. The elements themselves are no longer presupposed but rather deductively computed and deduced." As Cassirer notes in *Substance and function* and in *The Philosophy of Symbolic Forms*, Grassmann's geometry, during the nineteenth century, would return to and complete Leibniz's point of view in geometry by showing the fecundity of introducing the theory of groups in it. E. Cassirer, *Substance and Function*, pp. 96–97 (ECW 6, pp. 103–107) and E. Cassirer, *The Philosophy of Symbolic Forms*, vol. 3, p. 352 (ECW 13, p. 405).

principles of a formal science of order *to the quantities themselves*.[88] This, in Cassirer's view, is the very principle of infinitesimal calculus as it is conceived by Leibniz. Cassirer maintains, from this perspective, that infinitesimal calculus is not specific to Leibniz, even if it is more characteristic of Leibniz with respect to the quarrel with Newton regarding who first invented it. The premises of infinitesimal calculus had indeed been elaborated as early as the age of the preceding generation with Galileo, Cavalieri [1598–1647], Descartes, and Fermat [between 1600 and 1610–1665], but it is indeed with Leibniz that such calculus proved to be fully coherent with *general* epistemological principles concerning the notion of order and deduction.[89] If we take for example two series of which the magnitudes are in a determined relation, the relation remains even if the magnitudes considered are made infinitely small, to a point of rendering them inaccessible to intuition: from there stems the development of infinitesimal calculus. Also, this relation itself enables to determine the possibility for comparison, and, on the basis of *measurement*, between magnitudes regardless of their size.

Finally, this relational principle is applied by Leibniz to the field of physics because it is the only one which enables a mathematical study of the becoming. In contrast with the static viewpoint of Descartes which reduced the study of physics to that of equilibrium systems, in Leibniz's view, physics requires a dynamic viewpoint employing the notion of *force*.[90] Force is expressed at each instant of the *actual* state of a system's future, this actual state having the specific characteristic of carrying the integrality of its future states as well as its previous states. From this point of view, Leibniz is indeed one of the first having elevated to a status of principle the very notion of determinism: in his view, the rule for generating a numerical series obeys the same internal logic as the rule for generating the successive states of a physical system.[91] The mechanistic theory then allows a generalized calculus of events in nature and directly falls under the scope of the project for a "Universal Mathematics" in which what intuitively pertains to opposite pairs such as rest and motion become terms that are relative to the formal analysis of their rules of construction.

---

[88]While summarizing a stage in Leibniz's intellectual progression, Cassirer declares in E. Cassirer, *Das Erkenntnisproblem...*, Band 2, ECW 3, p. 126: "If the universal science would first of all limit itself to the reduction of all mental and real beings to numerical relationships, if it then learnt to discard the involvement of number and to understand all the presuppositions of the form purely from themselves: then the pure theory and the universal calculus of functions is now recognized as the real and deepest instrument for determining the number and the magnitude themselves."

[89]E. Cassirer, *Das Erkenntnisproblem...*, Band 2, ECW 3, p. 128: "Leibniz' achievement consists in the fact that he discovered a unified conceptual foundation for all these various sketches which in their implementation were limited to a specific domain."

[90]E. Cassirer, *The Philosophy of Symbolic Forms*, vol. 3, p. 456 (ECW 13, pp. 529–530).

[91]Commenting a letter by Varignon to Leibniz, Cassirer notes in E. Cassirer, *Das Erkenntnisproblem...*, Band 2, ECW 3, p. 132: "Each moment of the becoming must be derivable as distinctly following from the totality of the preceding moments and be preformed from them in their entire peculiarity."

1.3.3 The Role of Signs

As we have seen, Descartes' project for a universal language was confronted with the prerequisite constitution of "True Philosophy". But the development by Leibniz of samples of a general science in the sense of a universal theory of functional relations (and not only of primary elements) reformulated otherwise the problem of a universal language. Indeed, if the samples of the general science have succeeded in revealing the internal solidarity between a certain domain of objects and their corresponding written characters—the most manifest example of this being the invention of specific signs for differential and integral calculus[92]—it remains that the nature of the coupling between "general science" and "universal characteristic" in which it occurs appears, at first glance, to be problematic: it would indeed be necessary to *already* possess an adequate characteristic in order to have access to the general science, whereas it would be necessary to *already* have a general science in order to develop an adequate characteristic. The only means to overcome what appears to be a vicious circle consists in considering the role of signs with respect to knowledge in a different manner.

Henceforth, signs must be considered to not only be the vehicles of thought, but as indeed participating in its internal elaboration: this is why knowledge, in its algebraic deductive moment, can be *blind*—in the sense that it has recourse to signs and to signs only without the intuition of that to which they refer—without making it any less *adequate*. Also, signs not only have an *abbreviative* capacity, in the sense that they summarize within themselves an infinite number of elements, they also have a *prospective* capacity, inasmuch as, when *left to their own relations*, they lead to the discovery of new integrally determined relations.[93] It is in this respect that the vicious circle, for Leibniz, is overcome: it is a matter of *trusting signs*, even without the intuition of that to which they refer, so as to make them participate, in deductive progressions, in the adequate development of knowledge. Knowledge through

---

[92] E. Cassirer, *The Philosophy of Symbolic Forms*, vol. 1, pp. 85–86 (ECW 11, p.): "A clear understanding of the fundamental concepts of Galileo's mechanics became possible only when the universal logical locus of these concepts was, as it were, determined and a universally valid mathematical-logical sign for them was created in the algorism of the differential calculus. And then, taking as his point of departure the problems connected with the discovery of the analysis of infinity, Leibniz was soon able to formulate the universal problem inherent in the function of symbolism, and to raise his universal "characteristic" to a truly philosophical plane. In his view, the logic of things, i.e., of the material concepts and relations on which the structure of science rests, cannot be separated from the logic of signs. For the sign is no mere accidental cloak of the idea, but its necessary and essential organ. It serves not merely to communicate a complete and given thought-content, but is an instrument, by means of which this content develops and fully defines itself. The conceptual definition of a content goes hand in hand with its stabilization in some characteristic sign. Consequently, all truly strict and exact thought is sustained by the *symbolics* and *semiotics* on which it is based."

[93] E. Cassirer, *The Philosophy of Symbolic Forms*, vol. 1, p. 109 (ECW 11, p.): "It is one of the essential advantages of the sign—as Leibniz pointed out in his *Characteristica generalis*, that it serves not only to represent, but above all to *discover* certain logical relations—that it not only offers a symbolic abbreviation for what is already known, but opens up new roads into the unknown."

signs becomes, thereby, not a resembling while passive *image* of reality, but rather an *effective participation in its constitution*. It is precisely the reason for which Cassirer calls such knowledge *symbolic*: reality thenceforth manifests *directly* within knowledge as *order and connection between elements arranged in series*. Thus, the adjective "symbolic" acquires yet another aspect: it is no longer only associated with the establishment of relations between quantities in heterogeneous registers all pertaining nonetheless to the natural sciences as was the case with Galileo,[94] but it is thenceforth directly associated with the functional character of *any* knowledge within which the order and connection of things passes through the order and connection of signs. Arranging elements in series as it had been conceived by Galileo in the mathematical and physical fields acquires, after Leibniz, a directly algorithmic weight in the search for generalized functional correspondences. However, *there is no algorithm of which the domain of validity would be universal from the onset:* "general science" only manifests in the form of "samples" related to specific domains. Thus, the unlimited character of the number of objects arranged in series remains manageable because it falls within domain specificity.

Such a perspective has a radical consequence on the logical principle at first advocated by Leibniz according to which, in any proposition, predicates are "contained" within the subject. The samples of general science have shown, by their existence, that the relation between subject and predicates encompasses infinity: from then on, symbolic knowledge, by encompassing the possibility of an algebraic "blind" moment, that is, outside of any spatial intuition, does not submit to a logical relation of a container-to-content type which would only apply to cases consisting in the enumeration of a finite number of elements. On the other hand, the limitation is shifted from the number of objects arranged in series towards the *domain* of the studied object, either geometrical, arithmetic, or physical. It is a matter then of understanding not why such or such predicate is arbitrarily contained within such or such subject, but how such a series of predicates is associated with the subject within its own specific space and time.[95]

1.3.4 Primary Qualities and Extension of the Functionalist Perspective

The certitude of the knowledge produced by mathematical physics rested, since the beginning of classical science in the seventeenth century, upon a radical distinction between objectivity and subjectivity. This distinction was itself founded upon a

---

[94]E. Cassirer, *Das Erkenntnisproblem...*, Band 2, ECW 3, p. 137: "Ideas are not images but rather symbols of reality; they do not imitate all the particular traits and features of an objectively determined being but it is enough that they perfectly represent in themselves and so to say translate in their own language the relationships which exist between the particular elements of this being." Cf. *supra* the remarks of § 12.

[95]E. Cassirer, *Das Erkenntnisproblem...*, Band 2, ECW 3, p. 151: "It is not enough to know that the "character" B is forever contained in a universal concept A. We rather have to acknowledge the necessity according to which a certain particularity can be identified in a specific and unique "subject". Furthermore we have to understand why this particularity emerges at this precise moment and not earlier or later."

gnoseological distinction between *primary qualities* (of a mathematical nature) associated with necessity and *secondary qualities* (linked to sensible perception) which stemmed from contingency. This distinction enabled, in particular for Galileo, to circumscribe the perimeter of the certitude specific to mathematical physics and to distinguish it from any other certitude stemming from a different order, be it metaphysical or teleological. The list of these primary qualities immediately came to mind: whilst it was impossible to conceive of an object of the world without involving these primary qualities, it remained however possible to conceive of one without secondary qualities. These pertained to colors, smells, and in general to sense perception; they did not have an objective counterpart because their disappearance did not cause the object itself to disappear, even if its appearance could be altered. On the other hand, the primary qualities were necessary to the very conception of the object and were limited, in Galileo's view,[96] to properties pertaining to space, geometry, and number.[97]

However, the distinction between two sorts of qualities could not fail to raise the question of the grounding in reality of the objectivity specific to science. For Galileo, such grounding was conceived in the following manner: it was necessary to introduce the requirement of demonstration in order to reveal the coherence existing between the various phases of perception,[98] thereby assuming that nature was *in itself* interpretable in the form of geometric schemas. Introducing the necessity of geometry in the coherence of perception then imposed, in order to account for it, that principles anterior to any empirical experience be set. Thus, the *geometric readability* attributed to physical nature corresponded to its anticipatable *intelligibility* based on *a priori* principles. Physical events thereby stemmed from the bi-faceted domain of the sign, both legible and intelligible, but of which the legibility and the intelligibility was expressed, in this precise case, according to the very particular modality of geometry. This was eminently the case with the conservation of motion which, after Galileo, was

---

[96] The list of primary qualities was longer for Descartes who distinguished three fields of investigation, whether it was absolutely general or whether it was limited to bodies or to the soul. E. Cassirer, *Das Erkenntnisproblem...*, Band 1, ECW 2, p. 384: "There are different foundational concepts of this kind according to the different classes and problems that constitute the object of our research. Some of them such as Being, Number and Duration are evenly valid for all contents, whereas others such as Space, Figure and Movement concern more specifically the body, others again like the idea of Thinking, relate exclusively to the soul."

[97] E. Cassirer, *Das Erkenntnisproblem...*, Band 1, ECW 2, p. 326: "The matter or bodily substance is not comprehensible if one does not at the same time take its characters of limitation, of spatial figure and of magnitude into consideration. As far as its individual certitude is to be grasped, one needs to regard it as determined according to its local and temporal situation and its state of motion. All these viewpoints, that can be gathered together under the fundamental category of Number, Time and Space, belong therefore necessarily to its concept, from which they cannot be separated by the effort of the subjective 'imaginative force'." He even went so far as to place among secondary qualities the tactile sensation of resistance and, with it, the notion of weight which did not form part, in his view, of the concept of body. Cf. E. Cassirer, *Das Erkenntnisproblem...*, Band 1, ECW 2, pp. 327–328.

[98] E. Cassirer, *Das Erkenntnisproblem...*, Band 1, ECW 2, p. 331: "We reach this necessity by relating the observations to the ideal geometrical fundamental schemas and impressing on observations the logical form of these schemas."

elevated to a status of principle under the name of the principle of inertia: its geometric legibility manifested itself in the *rectilinear* and *uniform* character attributed to the concept of motion; its intelligibility consisted in supposing *a priori* that any velocity of an object of the world fell under its jurisdiction.[99]

Two questions were unavoidable concerning the nature of primary qualities. On the one hand, there was a methodological question: how are their reciprocal relations to be envisioned? On the other hand, there was an epistemological question: where did their explanatory power proceed from? It was necessary, in order to answer the methodological question, to study the relations between numbers, space, and geometry; and, regarding the epistemological question, to ask upon what founded the certitude of mathematical physics.

According to Cassirer, among the three Galilean primary qualities—number, space, geometry—it is the notion of number which was the least controversial in the historical development of modern physics. Indeed, the notion of number, taken in isolation, had been progressively replaced by the notion of mathematical function and the two presentations of infinitesimal calculus (Leibnizian infinitesimal calculus and Newtonian method of fluxions) marked the peak of this shift.[100] The case of the two other primary qualities was more delicate because the functional transformation[101] which would progressively affect them would require going through lengthy metaphysical controversies concerning the status of space and geometry, both notions being mutually related.

In the case of space, it was the status of the principle of inertia which posed the first difficulties. Indeed, the principle of inertia required referring to an *absolute* space in order for the notion of motion to be meaningful: thus, the Galilean relativity of place, indispensable in founding modern physics,[102] was nevertheless required to confer to space itself a non-relative status, unless one is to give up the very idea of

---

[99] E. Cassirer, *Das Erkenntnisproblem...*, Band 1, ECW 2, p. 333: "Hence modern knowledge does not conclude from the quality and composition of an isolated form of movement to its persistence and duration but, on the contrary, it is through the universal law of conservation that speed becomes a reality in the sense of knowledge, a permanent and determined 'essence', which only appears in the immediate, naive conception as an arbitrary emerging and disappearing content."

[100] Cf. § 132.

[101] As we have already noted in the introduction to this chapter, a distinction must be made between the strict sense of function pertaining to the mathematical case as first developed by Leibniz and the generalization of the idea of function to a broader perimeter of knowledge as it is viewed by Cassirer: It is in the second sense that the notion of function is employed here, a sense which enables to account for the generalization of the functional point of view in physics by the progressive conversion of the three primary qualities to this point of view.

[102] E. Cassirer, *Einstein's Theory of Relativity*, p. 361 (ECW 10, p.): "Galileo's doctrine of motion is rooted in nothing less and nothing more than in the choice of a new standpoint from which to estimate and measure the phenomena of motion in the universe. By this choice, there was given him at once the law of inertia and in it the real foundation of the new view of nature. The ancient view saw in place a certain physical property that produced definite physical effects. The "here" and "there", the "above" and "below", were for it no mere relations; but the particular point of space was taken as an independent real, which consequently was provided with particular forces."

motion, as Euler [1707–1783] had accurately seen.[103] There were therefore reasons to suppose principles which could, in no case, manifest within experience such as it appeared in mathematical physics all the while being indispensable to its constitution, contrarily to the methodological position with which Newton wished to keep.[104] But, *a contrario*, if mathematical physics was to be something other than simple contingent ramblings concerning nature, it was necessary to cease conferring absolute reality to space, as Euler pointed out to the proponents of the Leibnizian school who considered space in a functional manner as a simple system of intelligible relations.[105] It is this paradox which regularly rekindled, throughout the whole eighteenth century, the already ancient quarrel between the Newtonian conception of space interpreted as absolute space and the Leibnizian conception of space only conceived as a system of relations. This paradox concerning the status of space went hand in hand with a paradox concerning the status of the last primary quality described by Galileo, geometry.

The status of geometry, and more broadly of mathematics, was directly linked to the degree of *reality* one wished to confer to mathematical physics. Should the coherence between the phases of experience be associated with the geometrical certitude of proof because geometry was itself of an empirical nature as was experience, or because, on the contrary, it was of an *a priori* nature? As Cassirer noted, the debates surrounding this matter questioned the Galilean divide between primary and secondary qualities: the empirical conception had indeed for consequence to maintain

---

[103] E. Cassirer, *Das Erkenntnisproblem...*, Band 2, ECW 3, pp. 407–408: "Once again it is particularly the law of inertia which imposes on us the idea of absolute space and time. We are in that case presented with a most difficult paradox: what the experience denies us and for ever excludes from its sphere seems precisely to be these laws the experience itself demands for its justification (latin quote in footnote from Euler's *Theoria motus corporum solidorum seu rigidorum ex primis nostrae cognitionis principiis stabilita*, chap. II, § 81: "Anyone who would deny the absoluteness of space will fall into the greatest difficulties. Because he must indeed reject both absolute movement and absolute rest as vain words deprived of any meaning, not only must he reject the laws of movement which rest on this principle [of inertia] but he will also be forced to claim that there are no such things as laws of movement.)"."

[104] Cassirer summarizes the problem as follows in E. Cassirer, *Das Erkenntnisproblem...*, Band 2, ECW 3, p. 391: "If the absolute space, although it can never be given to us in any way, is nevertheless described as the indispensable principles of mechanics, then it is erroneous to depict experience as the limit of the content of our knowledge—so a "metaphysical" concept is imbedded in the foundations of mathematical physics. But as a consequence the force of pure induction, as Newton has understood and revealed it, would be already broken. Within the methodological rules which he set up for research, the first request is to admit nothing else than 'true causes', i.e. such that they prove to be of value in clarifying the phenomena. In its true meaning, the existence of absolute space and time is no 'vera causa': no natural phenomenon can supply us with its safe knowledge, no experience can legitimate or refute it."

[105] E. Cassirer, *Das Erkenntnisproblem...*, Band 2, ECW 3, p. 394: "While the pure space and time are described as abstract concepts and a separate physical existence is refused to both of them, this actually assigns to them the highest rank in the system of knowledge."

that there was no reason to attribute more *reality* to primary qualities than to secondary quantities[106] and that primary qualities only represented an easier *subjective access*.[107] The loss by primary qualities of any specific status thus ended up depriving geometry of any specific status also.[108] If a special place was to be maintained for the anticipative power of geometry, it would be necessary to interpret otherwise its role with respect to knowledge. Yet, until then, the certitude conferred by physics to mathematics had been sought in an *external* foundation: the distinction between primary and secondary qualities had played this epistemological role all the while feeding the controversy around the status to be conferred to geometry. Yet, as noted by Euler, there was in the end a paradox in seeking elsewhere than in mathematical physics a more certain foundation whereas only mathematical physics satisfied the conditions of this certainty. In order to lift the paradox, it was necessary to *keep solely with mathematical physics* while avoiding having recourse to a divide between primary and secondary qualities which only cast aside the problem of the status of

---

[106]In this respect, Cassirer cites the 1752 letters by Maupertuis [1698–1759] which completely throw into question the distinction between primary and secondary qualities; E. Cassirer, *Das Erkenntnisproblem…*, Band 2, ECW 3, p. 409: "What drives us to believe that Extension, Figure and Movement are not only subjective sensations in us but rather correspond to an independent, self-sustained reality in the bodies themselves? The justifications that one traditionally invokes to support this claim do not in any way stand up to a closer analysis. "I touch a body and I get a sensation of rigidity which seems to adhere much harder than odor, sound and taste. […] However, focusing on what rigidity and extension are in them, I do not find any justification to adopt that they belong to another category than odor, sound and taste." (Letter #4)."

[107]E. Cassirer, *Das Erkenntnisproblem…*, Band 2, ECW 3, p. 356: "If we attribute to extension another sort of certitude and a higher form of "being" than color or any specific second quality, then this difference [between second and first quality] has no real meaning. It is rather to be conceived of as the mere expression of the fact that extension features an easier point of application of our knowledge, in so far as any of its part can be reached by the evenly addition of a segment that we place as a yardstick. By contrast, such an easy and obvious comparison of various figures is accessible to no other domain. The reason of the certitude of mathematics is only this "repeatability" (réplicabilité –*in French*–) of meaningful ideas from which it starts."

[108]This is for instance the case with Boscovich [1711–1787] for whom the ontologically differentiated nature of the extended points finishes by removing any possibility of a measurement common to places; E. Cassirer, *Das Erkenntnisproblem…*, Band 2, ECW 3, p. 436: "So far, the line is defined as the embodiment of points of space which in return are only known to us as qualities deriving from points of force. […]. The unitary segment suffers an inner change as it is drawn to another place of space where another assignment of physical points is to be found and therefore where from now on real "places" that are different from before are also recomposed. […]. We attend here a liquidation of geometry for it is the effective nature of bodies that decides of the value of mathematical concepts." Cassirer makes no mistake here: It is not a pre-Einsteinian idea of relativity which challenged the possibility of a measurement between two referentials, but only of a *relativism* aiming to deny geometry any anticipatory power.

the *a priori*[109] on the one hand and depreciated the *a priori* value of geometry on the other hand.[110]

This position, developed by Euler, had two capital consequences regarding the status of knowledge. First, mathematical physics became in itself that with respect to which it was possible to evaluate the degree of certainty of knowledge. From this standpoint, there was indeed an *epistemological extension of the perimeter of validity specific to mathematical physics* because it became that against which the totality of knowledge was to be evaluated.[111] Secondly, this implied conceiving otherwise the certitude of mathematical physics which stemmed neither from a presumed coherence of empirical matter nor from *a priori* principles: it was a matter of taking experience as a whole while keeping with a *geometrization of space-time* appertaining to all possible phenomena since it is from such geometrization that stemmed the certitude of the knowledge produced by mathematical physics. The foundation thence conferred to mathematical physics being self-sufficient, one could say, only made sense if geometrization was carried forward on an ever-increasing scale. The case of the principle of inertia illustrated this new epistemological attitude: on the one hand, it was finally no longer space and time as such which were the objects of scientific investigation, but the way in which these notions were used with respect to the principle of inertia[112]; On the other hand, the principle itself only acquires the sense of a foundation when it can extend to yet unexplored phenomena, such as those pertaining to chemistry where the notion of quantitative conservation definitely takes precedence over a qualitative conception of substances.[113] This geometrization of space-time already immerses us within a quasi-Kantian point of view: as noted by

[109] E. Cassirer, *Das Erkenntnisproblem...*, Band 2, ECW 3, p. 424: "Regardless of how this division is being made and how we differentiate the "ideal" from the "real", one point has to be established from the very beginning: the exact concepts of mathematics and the concrete objects of nature belong to the same domains of knowledge."

[110] E. Cassirer, *Das Erkenntnisproblem...*, Band 2, ECW 3, p. 424: "The fundamental concept of the new analysis [infinitesimal calculus] blows up all the usual divisions of ontology. What meaning can the old division of knowledge into different, separate "capacities" still have if the most important and most certain content of knowledge escapes it?" If one understands this concept in the sense that the philosophy of the school [i.e. [Wolff's] gives, mathematics does not belong to the "pure intellect."

[111] E. Cassirer, *Das Erkenntnisproblem...*, Band 2, ECW 3, p. 402: "With the doctrine of Euler the new mathematical science reaches a degree of maturity from which it henceforth sets to establish the real standard of "objectivity" from itself, instead of letting it be imposed by some foreign interest."

[112] E. Cassirer, *Das Erkenntnisproblem...*, Band 2, ECW 3, p. 401: "What makes the whole decisive question is not what space and time are for themselves and in themselves, but rather how they are applied in the articulation and formulation of the inertia principle."

[113] E. Cassirer, *Das Erkenntnisproblem...*, Band 2, ECW 3, pp. 367–368: "The logical parallel which exists here between the foundations of mechanics and that of chemistry is particularly symptomatic and informative of the evolution of the general concept of knowledge. Alongside the concept of substance is that of causality, through which this analogy proves its worth. Likewise, the physical concept of cause first had to be unfolded by means of the concept of magnitude so that it could reach the full determination of its meaning. [...]. Consequently, the new discipline [of chemistry] fits perfectly in the frame of the physicalist method Newton had elaborated for the knowledge of nature in general."

Cassirer when revisiting in a less polemical manner the suggestion put forth by Schopenhauer,[114] it is more the state of the sciences at the end of the eighteenth century than the Kantian doctrine as such which accounts for the *ideality* conferred to space and to time, that is, there is a shift in the idea of necessity from the logical sphere of proof towards the physico-mathematical sphere of geometry.

1.4   The Geometry of Space-Time as the Originary Semiotic Locus of Modern Science

Four points thus emerged, in Cassirer's view, from the epistemology of the natural sciences during the second half of the eighteenth century, in particular in the Franco-German setting of the Academy of Berlin. These four points were divided into two sets, the first of which was not an object of controversy, contrarily to the second. In the first case, on the one hand, there was consensus regarding the fact that mathematical physics indeed said something about reality and that it did not only consist in simple words devoid of an ontological stabilizer and, on the other hand, that mathematical physics *was its own norm* without having to compare its own value to that of other modes of discourse, as had been the case in the seventeenth century during which such a comparison made up the very substance of its epistemological concerns. The second set was, for its part, a matter for controversy: confronted with the powerlessness of logical laws (essentially the principles of non-contradiction and of syllogism inherited from Aristotle's logic) in justifying the causal necessity being deployed in the natural sciences, it was to be concluded either that causal necessity *did not actually exist*, following the path taken by the empiricist thesis defended in particular by Hume and Maupertuis,[115] or that in the natural sciences it came from *another source* than the logical laws in question[116] and that this necessity had to be put into relation with the *geometry of space-time* as the locus where meaningful relations between objects and concepts became *expressible*.

It is this latter option to which Cassirer gave particularly close attention inasmuch as it intersected with his own epistemological choices.[117]

---

[114]E. Cassirer, *Das Erkenntnisproblem...*, Band 2, ECW 3, p. 411: "The historical observation teaches and immediately proves that Kant certainly didn't need to "discover" the doctrine of the ideality of space and time."

[115]E. Cassirer, *Das Erkenntnisproblem...*, Band 2, ECW 3, p. 355: "The first reception and continuation of the Humian way of thinking was found in the person of mathematician Maupertuis, who, while he would publish his writings in the Proceedings of the Berlin Academy, was the first to bring the Humian problem in the horizon of German philosophy."

[116]Before being adopted and explored in greater depths by Kant, this thesis was first defended by Christian Crusius [1715–1775]; Cassirer notes regarding the latter in E. Cassirer, *Das Erkenntnisproblem...*, Band 2, ECW 3, p. 500 footnote 16: "In contrast, the essential outcome of his [Crusius'] theory of knowledge lies in the insight that our causal conclusions require an independent principle different from that of contradiction and a different way of founding its certitude."

[117]From this point of view, his choices have complex relations with Kantian epistemology. As previously remarked, Kant occupies a distinct position within the broad picture Cassirer draws of

1.4.1 Beyond Logical Necessity

The issue of the necessity of physical laws as it appeared in the epistemological context of the second half of the eighteenth century can be summarized as follows: not only is the *logical* principle of non-contradiction inapt in justifying the physical necessity of the relation between a cause and an effect but *experience* is no longer the solution to the problem of causality, as opposed to the way in which the problem was approached since Antiquity.[118] Indeed, during that period, there was an attachment to avoiding the skepticism entailed by the difficulty of conceiving the connection between cause and effect by having recourse to psychology and to the notion of habit. But, as Hume had for the first time radically highlighted, it was thenceforth the anticipation of a *future* connection which risked having devastating effects[119] in the context of a reflection on the necessity of the laws of mathematical physics, and the role accorded by the Ancients to habit no longer enabled to resolve the problem. So the question arises as to where this *anticipative* necessity could come from, if anticipation there was.

For Cassirer, avoiding the return to skepticism involved a recourse to a *new kind of imagination* expressed using a new style, one which could be found particularly in Kant's *Dreams of a Spirit-Seer*, published in 1766. In this work, Kant showed that in basing oneself solely on the principle of non-contradiction, there would be no means to distinguish dreams from reality. Now, since this difference did indeed exist, it meant that there was, in experience, a way in which to operate it: it is the *functional continuity* of the relations between objects within experience that enables, in his view, to justify this difference. Progressively, in the elaboration of a response to the issue of the nature of experience, the idea appeared according to which necessity should not be conceived on the basis of the difference between contingent physical matter

---

philosophy and of modern science in *The Problem of Knowledge...* given that Kant serves, for Cassirer, as his own entry point into philosophy via the neo-Kantian philosophers of the school of Marburg whose teachings he followed. Also, it is with respect to Kant that Cassirer positioned his own epistemological work. However, all the while including within the perimeter of Kantian epistemology the post-Kantian advances in the field of the natural sciences during the 19th and twentieth centuries, he also distinguished himself from Kant when he gave a *semiotic* twist to this epistemology, as shown in the preceding pages.

[118] E. Cassirer, *Das Erkenntnisproblem...*, Band 2, ECW 3, p. 298: "Until then, Hume still moves on well-known tracks: if it were the core content of his doctrine, he would in fact not have been in any way further than the doctrine of experience of the ancient skepticism. The fact that cause and effect are not united through a conceptual relationship, the fact that the two of them can only produce an association in representation through the regular connection that binds them in experience, this idea wasn't only occasionally touched upon, rather it had evolved in a complete and coherent theory."

[119] E. Cassirer, *Das Erkenntnisproblem...*, Band 2, ECW 3, p. 299: "The experience which was till then considered as a panacea in which research would find comfort has now become an insoluble problem. Its value is no longer naively presupposed but has rather become a true riddle. The justification of our causal conclusions can neither be sought in logical conclusion nor in experience. Consequently, it is its very foundation which is now in question. It can be understood that we are endowed with a power to remember the past cases that we have perceived so that we can recall them again; but how we may have an overall look and are able to determine from our previous limited particular observations the totality of future events remains completely incomprehensible."

and logical conceptual necessity. This called for a reorganization of the distinction between necessity and contingency on the one hand and between experience and abstraction on the other. By starting with the hypothesis that the distinctions in question henceforth operated with respect to *relations* expressible through propositions (be these relations between cause and effect or between subject and predicate), and that some necessary propositions nevertheless pertained to mathematical physics and not to logic alone, the conclusion to draw should be that mathematics and physics *shared a same necessity* which pertained to *space-time* such as it had been theorized by Galileo, Descartes, and onwards. This change in perspective was considerable since it placed physics and mathematics on the *same* side, these deploying a same necessity within experience, henceforth understood as a measurable spatio-temporal environment distinguishable from material exteriority.

In doing this, there was no longer a need for attempting to justify a necessity inscribed within "things" and which would only be conceptually accessible through logic: what one would relate to *intuition* did precisely not pertain to *inference*.[120] As Euler had himself declared, the physicist assumes the existence of objects in nature and attributes reality to the functional relations established between these objects without having the need to justify their existence, deemed to belong to the category of the given and functionally accessible. Logic therefore no longer plays, in this epistemological configuration, the exclusive metaphysical role of a receptacle for the essence of things which knowledge would need to conceptually bring to light, as it had played until then.

1.4.2 The Geometrization of Experience

One can thence understand the complex part played by geometry in this epistemological apparatus: geometry, by virtue of the role played within it by the notion of demonstration, bears a relationship with logic but, by virtue of the forms it constructs, also bears a relationship with sensible space by involving physico-mathematical necessity within such space. However, a difficulty arises here: what allows to think that all spatio-temporal relations are expressible in geometric form? In fact, this is precisely not the case; there is, in sensible space, regulated relations *which are not of a logical nature all the while being geometrical*, as proven by the example of objects that are identical in their shape but which do not coincide in space, that is, objects which are "non-congruent", the status of which Kant had begun to ponder.[121] In these

---

[120] E. Cassirer, *Das Erkenntnisproblem...*, Band 2, ECW 3, p. 609: "The empirical reality is called "immediate" in so far as it is not necessary, in order to make it certain, to overcome consciousness and use a completely different way of being; but it is clear that, from a logical point of view, it must be seen as mediated by the conditions of thought as well as by pure intuition."

[121] E. Cassirer, E. Cassirer, *Das Erkenntnisproblem...*, Band 2, ECW 3, p. 518: "The particular and specific character which makes the difference between the right and left hand, lies neither in any property of the two hands themselves nor in the relationship between their individual parts; we must rather evaluate the two bodies against the whole of space, as a surveyor would think for himself."

conditions, how can geometry be conferred a role of mediator between intuition and concepts if it does not cover the full extent of spatial relations?

To answer this question, two points must be considered: on the one hand, it is indeed *with respect to geometry* that spatial relations such as non-congruence can be qualified as "non-logical" and, on the other hand, it is up to geometry to attempt to give them expression within its own domain.[122] There is thus at the foundation of the role played by geometry a *requirement for geometrization* which aims to make all relations present in sensible space expressible in geometrical terms.[123] But from where does it stem that this anticipated geometry of possible spatial relations is to be expressed as a *requirement*? The response elaborated by Kant consists in saying that requirements form the specific mode of the intervention of the subject within knowledge: by producing a law without simply obeying it,[124] the subject reconfigures the world in its totality for an end that the subject alone is capable of conceiving, that is, the anticipation of the legality of experience.

For Cassirer who radicalizes a point of view that had barely been suggested by Kant in the third *Critique*,[125] such anticipation only makes sense by means of language which plays the role of condition of possibility with respect to this anticipated legality. It is indeed language which enables to describe causal relations as necessary since it enables to regard causality as applicable to the *totality* of experience and, thereby, as unable to not take place.[126] As a vector of meaning, language *alone is susceptible* of

---

[122]Historically, this will actually be the case given that Legendre [1752–1833] will manage to describe non-congruence in a geometrical framework as early as 1794, that is, only a few years following Kant's work on this topic in 1768. Cf. Kant, "Von dem ersten Grunde des Unterschieds der Gegenden im Raume" [1768] in E. Cassirer (ed.), *Immanuel Kants Werke*, Band 2 pp. 393–400 and Legendre, *Eléments de géométrie*, livre VI, proposition XXV, p. 209.

[123]It is therefore by conforming to the *limits* of geometry that any potential reorganization of the physico-mathematical field unfolds: the history of mathematical physics and of the transformation of its geometrical models.

[124]E. Cassirer, *The Philosophy of Symbolic Forms*, vol. 1, p. 200 (ECW 11, pp. 149–150): "For Kant the concepts of the pure understanding can be applied to sensory intuitions only though the mediation of a third term, in which the two, although totally dissimilar, must come together—and he finds this mediation in the "transcendental schema", which is both intellectual and sensory. [...]. Language possesses such a "schema"—to which it must refer all intellectual representations before they can be sensuously apprehended and represented—in its terms for spatial contents and relations. [...]. The relations of "together", "side by side", "separate" provide it with a means of representing the most diverse qualitative relations, dependencies and oppositions."

[125]In the *Critique of Judgement*, § 59, Kant outlines a reflection on the role of language in the characterization of symbolic intuition: "Our language is full of indirect presentations of this sort, in which the expression does not contain the proper schema for the concept, but merely a symbol for reflection. Thus the words ground (support, basis), to depend (to be held up from above), to flow from something (instead of, to follow), substance (as Locke expresses it, the support of accidents), and countless others, are not schematically but symbolical hypotheses and expressions for concepts, not by means of a direct intuition, but only by analogy with it, i.e. by the transference of reflection upon an object of intuition to a quite different concept to which perhaps an intuition can never directly correspond." (*Critique of Judgement*, translated by J. H. Bernard, MacMillan & Co., London, 1914).

[126]E. Cassirer, *Substance and Function*, p. 258 (ECW 6, pp. 278–279): "The inductions of Kepler on planetary motion express only generalized "rules" of process, while the fundamental law, on which they rest, was first formulated in Newton's theory of gravitation. In Newton's theory we find

straying from the here and now and of presenting the world as a totality regulated by laws. There is therefore, in language, a *reservoir* of possible objectivity, susceptible of expressing necessity such as it manifests in space-time regarded as a whole, in particular in the form of causality. Conceiving in this manner the notion of necessity as the anticipation of possible objectivity pertains to a semiotic experience touching the totality of possible experiences having language for a vector. The requirement of the geometrization of space supposes, from this standpoint, that it is expressible in the form of propositions.

1.4.3 The Critique of Reason as Functional Philosophy

It then became clear that the reconfiguration of the notion of certitude from mathematical physics such as it had been envisioned first by Euler also had capital philosophical consequences regarding the very task of philosophy. It indeed no longer aimed to uncover an ontology only obtainable through concepts, as had been the case until then since Antiquity. It was thenceforth a matter of taking the full measure of the dissemination of the functional viewpoint towards philosophy and of dismissing as vain any research concerning the ultimate foundations of concepts: the task of philosophy would consist from that moment onwards in reflecting on the modalities of the construction of experience, that is, of determining how, starting with matter, a functional form of experience constitutes itself. It is in this sense that philosophy, for Kant, adopted regarding its own concepts the same attitude as Newton did regarding those of mathematical physics.[127]

2 The Symbolic Turn of Epistemology Following Newton

For Cassirer, Kant's work not only represents the pinnacle of the eighteenth century reflection regarding the natural sciences, but also the means by which to interpret their ulterior developments: transcendental philosophy thus becomes a true method

---

the ellipse not merely as a real form on the orbit of Mars, but we survey at a glance the whole of "possible" orbits. The Newtonian concept of a centripetal force, that diminishes according to the square of the distance, leads to a perfect disjunction of the empirical cases in general. The transition between these cases is henceforth exactly predetermined […]. Thus the "law" of gravitation contains in itself the field of facts, which it rules, and ascribes strict division to its field; while the merely empirical rule of planetary motion allows the particular cases to stand in loose conjunction without sharp delimitation."

[127] E. Cassirer, *Das Erkenntnisproblem…*, Band 2, ECW 3, p. 617: "What we can crave to know, is only on which path and in virtue of which condition the perception, starting from the mere "matter", constitutes itself as the scientific "form" of experience. Therefore, we do not have to seek for long where the experience stems from, but we ask what it is according to its pure logical structure. The fact that the true method of metaphysics is fundamentally the same as the one Newton has followed in the natural sciences, had already been underlined in the pre-critical writing *Über die Deutlichkeit der Grundsätze der natürlichen Theologie und der Moral* [E. Kant, Inquiry into the Distinctness of the Principles of Natural Theology and Morals]."

of investigation which can be applied to materials to which Kant could not have at his disposal during his life time. From this point of view, the epistemological framework established by Kant does not simply represent for Cassirer a bygone stage of the history of philosophy, but it is most of all that owing to which the modern natural sciences, as opposed to those of Antiquity and of the Middle Ages, become interpretable, *including after the Newtonian age of physics*. It is impossible to not reflect upon the reasons which led Cassirer to interpret the developments of physics after Newton—electricity, magnetism, thermodynamics, relativity, and even quantum mechanics—using the epistemological tools developed by Kant.

Would this be the result of a personal bias stemming from his own philosophical training marked by the legacy of the neo-Kantian Marburg school[128]? Or is there some true relevance in using Kantian epistemology in a context which Kant could not anticipate[129]?

2.1 Kantian Epistemology as Object and as Resource

The Marburg School endeavored to show that the epistemological tools developed by Kant had not become obsolete with the ulterior development of the natural sciences: these indeed possessed a profound unity in the issues they raised since the beginning of the modern era, which the epistemological tools developed by Kant—in particular

---

[128]Cassirer noted that the role played by the philosophy of Kant in Cohen's work had consisted in the constant reminder of the relevance of the transcendental point of view for accounting for the fact of science, irrespective of its content and historical development; E. Cassirer, "Hermann Cohen und die Erneuerung der Kantischen Philosophie" [1912], ECW 9, p. 122: "The "revolution in the way of thinking" which is carried out in the critique of reason is rooted in the transcendental way of posing a problem; according to Kant, "transcendental" means this mode of observation which does not as much starts from the objects than from the actual mode of knowing the objects.[...]. Consequently, the proper object of philosophy is not the "organization" of nature or of "psyche", but what only it has first to determine and discover, that is the "organization" of the knowledge of nature."

[129]It is also a legacy which he links, quite problematically, to Cohen; E. Cassirer, "Hermann Cohen und die Erneuerung der Kantischen Philosophie" [1912], ECW 9, pp. 127–128: "If it is true, as Cohen put it strongly, that "Kant couldn't be anyone else but a Newtonian", any transformation of Newtonian mechanics poses a threat to the very core of the system of "synthetic principles". However, the very development of Cohen's theory has refuted this viewpoint. Very energetically engaged as he was to put back at the center of attention the Newtonian system, he would follow, with the most decisive interest and unprejudiced acumen, the transformations that this system has experienced in the physics of the nineteenth century. He was one of the first who pointed out the philosophical meaning of Faraday's theory, he also explored the principles of Heinrich Herz' mechanics in order to understand them and ground their content in critical knowledge. Therefore, the orientation towards knowledge meant for him no connection to its temporal and contingent form. The philosopher recognizes that what is given in the mathematical sciences of nature means in the end that a problem is given."

his reflections regarding the geometry of space-time—had simply uncovered. It was therefore relevant to use Kantian epistemology in a post-Newtonian context.[130]

Cassirer's approach nevertheless distinguishes itself from that of the school of Marburg. It consists in appropriating a *new style* of thinking which appears at first glance to muddle the traditional Kantian points of reference by seemingly distancing itself from the transcendental method while nevertheless enabling new usages of Kantian epistemology. As strange as it may seem at first glance, it is with Goethe that Cassirer found this new style, as radically opposed as he may have been to the Kantian idea of a mathematized physics in the manner of Newton.[131] What is there in Goethe's work which may have enabled Cassirer to operate an extension of Kantian epistemology beyond its Newtonian limits?

Goethe was in search of a whole other physics than that of Newton, one which would have been of a directly *morphological* nature. His morphological physics would have had for ambition to study the *generative* power of the forms of nature, the fecundity of this generation not being essentially attributed to calculatory processes as is the case in mathematical physics,[132] but rather to that which makes possible the *individuality* of forms. At the foundation of this morphological physics, Goethe presents a new usage of the term *symbol*, thenceforth conceived as the *deployment of a form* from a primitive Idea[133] and not, as per the strictly Kantian point of view,

---

[130]Cassirer remarks concerning the epistemological position of Hermann Cohen in E. Cassirer, "Hermann Cohen und die Erneuerung der Kantischen Philosophie" [1912], ECW 9, pp. 127–128: "If it is true, as Cohen put it strongly, that "Kant couldn't be anyone else but a Newtonian", any transformation of Newtonian mechanics poses a threat to the very core of the system of "synthetic principles". However, the very development of Cohen's theory has refuted this viewpoint. Very energetically engaged as he was to put back at the center of attention the Newtonian system, he would follow, with the most decisive interest and unprejudiced acumen, the transformations that this system has experienced in the physics of the nineteenth century. He was one of the first who pointed out the philosophical meaning of Faraday's theory, he also explored the principles of Heinrich Herz' mechanics in order to understand them and ground their content in critical knowledge. Therefore, the orientation towards knowledge meant for him no connection to its temporal and contingent form. The philosopher recognizes that what is given in the mathematical sciences of nature means in the end that a problem is given."

[131]E. Cassirer, "Goethe and the Kantian Philosophy" in *Rousseau, Kant, Goethe, Two Essays*, ECW 24, p. 543: "Kant demanded that mathematics should enter into every part of the theory of nature, Goethe energetically rejected any such notion."

[132]E. Cassirer, „Goethe und die mathematische Physik. Eine erkenntnistheoretische Betrachtung", ECW 9, p. 291: "But the reduction to a numerical value leaves nothing from the particular empirical intuition which is the qualitative starting point of physical observation. The initially qualitative values are sought to be transformed in pure positional values which are characterized by nothing else than connections to the other terms of the series. All individuality of intuition is lost in such connections; all that manifests itself initially as an independent substrate is only taken into consideration by exact research in so far as it is progressively transplanted in a system of relationships and functional dependency."

[133]Goethe had embarked upon the quest to find the "primitive plant" from which the form "plant" had deployed itself in nature. Cf. E. Cassirer, "Goethe and the Kantian Philosophy" in *Rousseau, Kant, Goethe, Two Essays*, ECW 24, p. 555: "Goethe later learned to think of the original plant in a different fashion. He no longer hoped to see it with his eyes and to grasp it with his hands. But the value of his theory did not seem to him to have diminished or to have been called into question on

as the presentation of an only reflexive organization of nature. It is in this sense that there is, in Cassirer's undertaking, a turn which can be qualified as *aesthetic*: not that he would attempt to exclusively concentrate his attention on the phenomenon of art or on the very particular Kantian acceptation of the term "aesthetic" interpreted as that which makes possible conceptualization at the very level of sense perception; the term "aesthetic" must be understood here as a sign that the question of the organization and of the deployment of forms becomes central from both a standpoint internal to science and from an external standpoint. For Goethe, it is in the sense where forms are only intelligible as modes of organization and diffusion that they are said to be "symbolic". This recourse to Goethe,[134] which distinguishes Cassirer from his mentors at Marburg, has two immediate consequences regarding the usage which can be made of Kantian epistemology after Newton: if it is indeed, as we shall see, a matter of grasping the deployment of "symbolic forms", then this deployment can be examined in modern physics starting with its seventeenth century roots, but also *after* the Newtonian age, up until the most contemporary period as well as *elsewhere* than in the natural sciences if one makes the hypothesis that any manifestation of meaning is analyzable in terms of organization, diffusion, and competition of forms.[135] The meaning attached to the adjective "symbolic" now authorizes this double progression, in the epistemology of physics, but also beyond it.

The epistemological attitude to adopt regarding physics and mathematics changes considerably with this "aesthetic" turning point in Cassirer's thought. From the standpoint of physics, it is not a matter of returning to a pre-Eulerian point of view which would seek elsewhere than in mathematical physics a means to evaluate its certitude. Physics, in its own manner, continues to deploy its form and the very particular modality of its necessity without having to rely on anything else than itself. But the deployment of this specific form which is physics is no less *regional* when one adopts a morphological point of view where the organization and mutual relations of forms of objectification prevail in culture as a whole because there is no reason to suppose *a priori* that the apparition and diffusion of forms of objectification are organically linked to mathematics, unless one were to return to a substantial conception in which the space-time of physics would be ontologically first. Now, this is precisely not the case for Cassirer, as his whole historical reconstruction of the development of

---

that account. Now he no longer took offense when the original plant was called ideal. He himself called it that, and he used another expression that is genuinely Goethean and profoundly significant. He called it a *symbol*."

[134] E. Cassirer, *The Platonic Renaissance in England*, translated by James P. Pettegrove, Nelson, 1953, pp. 199–200 (ECW 14: 378–379): "Goethe is indebted to Shaftesbury for the concept of 'inward form' which permeates his reflections on nature as well as on art and his early fragment, 'Die Natur', expresses the same feeling that inspires Shaftesbury's apostrophe to nature in 'The Moralists'.

[135] E. Cassirer, *The Philosophy of Symbolic Forms*, vol. 1, p. 81 (ECW 11, pp. 10–11): "[…] in the course of its development every basic cultural form tends to represent itself not as a part but as a whole, laying claim to an absolute and not merely relative validity, not contenting itself with its special sphere, but seeking to imprint its own characteristic stamp on the whole realm of being and the whole life of the spirit. From this striving toward the absolute inherent in each special sphere arise the conflicts of culture and antinomies within the concept of culture."

modern physics since the Renaissance has shown: it is a new attitude concerning the symbolic institution of the *signs* of language which progressively generated the functionalization of the space-time of physics. In this perspective, the role of mathematics finds itself, at the same time, redefined: if mathematics indeed plays a constitutive role in the deployment of the symbolic forms specific to the natural sciences—as Kant and the Marburg School rightly emphasized so strongly—it does not however bear an *organic* relation with what presents as objects in other symbolic forms. Also, although mathematics can indeed *intervene* in these other forms, it does not necessarily play a constitutive role in them as such, precisely because it is confronted with *instituted* objects, such as myths or languages, rather than with objects needed to be *constituted*, such as mathematical or physical objects.

These two consequences, internal and external to the natural sciences, form the guidelines of the epistemological project of a mature Cassirer who, one the one hand, continued during his whole life to study the evolution of modern physics and who, on the other hand, extended his study of the organization of other symbolic forms beyond the strict field of the epistemology of the natural sciences.

The Kantian transcendental method is thus endowed by Cassirer with an extension which could be qualified as "conservative" inasmuch as it does not break with the Marburg school,[136] contrarily to attitudes in favor of rupture as advocated for instance by Heidegger in his debate with Cassirer during their 1929 encounter in Davos. In Cassirer's view, Kant and his Marburgian posterity definitely dismissed the idea defended by Heidegger according to which the task of the philosopher consists in reflecting upon the foundation of ontology, precisely because the very notion of ontological foundation has been replaced by a multiplicity of symbolic forms of which it is necessary to consider, from a philosophical standpoint, the relation to objectification specific to each.[137]

---

[136]E. Cassirer, in *European Existentialism*, ed. Nino Langiulli, New Brunswick, N.J.: Transaction, 1997, pp. 202–203 (ECN 17, pp. 108–109): "I hold to the Kantian formulation of the question of the transcendental. The essential method of the transcendental lies in this, that it begins with a given. This I inquire into the possibility of the given called 'language'."

[137]E. Cassirer, in *European Existentialism*, ed. Nino Langiulli, New Brunswick, N.J.: Transaction, 1997, p. 202 (ECN 17, pp. 108–109): "However, here an essential difference seems to me to obtain, which is in fact what Kant called the Copernican revolution. […]. What is new in this revolution seems to lie in this, that there is now no longer a single such structure of being but rather that we have completely different structures of being. Each new structure of being has new a priori presuppositions. Kant shows how every kind of new Form always bears upon a new world of objectivities. In that way a whole new multiplicity enters into the problem of the object as such. By that means the old dogmatic metaphysics becomes the new Kantian metaphysics. The being of the old metaphysics was substance, that *one* which underlies. In the new metaphysics, being is in my language no longer the being of a substance, but the being that proceeds from a manifold of functional determinations and meanings."

2.2   The Epistemology Internal to the Natural Sciences Following Newton

For the time being, it is fitting to begin by studying the first aspect of the conservative extension of Kantian epistemology proposed by Cassirer, the aspect which concerns that which lies beyond Newtonianism in the natural sciences. For Cassirer, this study must concern less the ulterior content of the natural sciences than the modalities of their specific forms, of which the evolution, within the functional approach, essentially pertains to the interpretation of space.

Cassirer distinguishes three stages in the dissemination of the functional approach, of which only the last concerns the post-Newtonian extension of the natural sciences, which Cassirer calls the physics of "principles".[138] Here follows a brief review of the first two of these stages.

2.2.1   From "Substantial Forms" to "Serial Forms"

The first stage corresponds, as seen earlier, to ancient and medieval theories which interpret the nature of the concept on the basis of the difference between the particular and the general: is thus conceived as properly conceptual the common substance which enables to trace in both directions the movement between particular and general terms. This difference is doubled, at the same time, by a similitude enabling to conceive of continuity between the terms. But this similitude is not only subjective because substance has for common feature with respect to all objects that it not only allows the continuity of their perception in space and time, but also the interpretation of this continuity in both general and natural terms, according to the causality and finality which concretely move the objects in question and which reciprocally associate them.[139] It is for this reason that the substantial viewpoint involves in itself that the concept anticipates the nature of objects as they are perceived in space according to the modality of an interlocking between genii and species which reveals itself first and foremost in the organization of living matter.[140] There is a physics of *resemblance* between concepts and sensibility within perceived space. The scientific revolution

---

[138]E. Cassirer, *The Philosophy of Symbolic Forms*, vol. 3, p. 460 (ECW 13, p. 534): "In its general structure nineteenth-century physics might be characterized as a physics not of images and models but of principles."

[139]E. Cassirer, *Substance and Function*, p. 7 (ECW 6, pp. 5–6): "The selection of what is common remains an empty play of ideas if it is not assumed that what is gained is, at the same time, the real *Form* which guarantees the causal and teleological connection of particular things. The real and ultimate similarities of things are also the creative forces from which they spring and according to which they are formed."

[140]E. Cassirer, *Substance and Function*, p. 7 (ECW 6, p. 6): "The biological species signifies both the end toward which the living individual strives and the immanent force by which evolution is guided. The logical doctrine of construction of the concept and of definition can only be built up with reference to these fundamental relations of the real."

which begins in the Renaissance and which makes possible the appearance of mathematical physics during the seventeenth century constitutes the second stage and consists precisely in foregoing a theory of resemblance between concepts and the sensible.[141] The opposition drawn by Cassirer between substance in ancient science and function in modern science proceeds indeed from a difference pertaining to the nature of the concept.

2.2.2   From "Serial Forms" to "Symbolic Forms"

Modern theory, for its part, interprets the nature of concepts according to the schema of systematic arrangement into series which is not limited to continuity through similarity but which can involve any sort of possible relations in such arrangements.[142] From a logical standpoint, it is therefore the notion of relation rather than of term which finds itself placed at the center of the knowledge-producing apparatus: it is no longer a matter of neglecting particular aspects of objects so as to note only the similarities they share with other objects but rather to uncover the rule of object production in function of the chosen serial principle.[143] No longer is it a matter of removing the intrinsic particularities from objects but rather of situating them within

---

[141] E. Cassirer, *The Philosophy of Symbolic Forms*, vol. 3, p. 452 (ECW 13, p. 524): "But even the founders of classical mechanics, Galileo and Kepler, Huyghens and Newton, stand only at the beginning not at the end of this development. Their achievement consists essentially in taking the step from empirical intuition to pure intuition: in taking the world not as a manifold of perceptions but as a manifold of forms, figures, and magnitudes. But this figurative synthesis still preserves a certain limitation: it is restricted to the datum of *pure space*. It is pure space that serves as model and schema for the building of all the geometrical and mechanical models to which classical physics reduces the multiplicity of empirical phenomena and in which it sees the prototype of all scientific explanation of nature."

[142] E. Cassirer, *Substance and Function*, p. 16 (ECW 6, p. 15): "In truth, it will be seen that a series of contents in this conceptual ordering may be arranged according to the most divergent points of view; […]. We may place series in which between each number and the succeeding member there prevails a certain degree of difference. Thus we can conceive members of series ordered according to equality or inequality, number and magnitude, spatial and temporal relations, or causal dependence. The *relation of necessity* thus produced is in each case decisive; the concept is merely the expression and husk of it, and is not the generic presentation, which may arise incidentally under special circumstances, but which does not enter as an effective element into the definition of the concept."

[143] E. Cassirer, *Substance and Function*, p. 19 (ECW 6, p. 19): "When a mathematician makes his formula more general, this means not only that he is to retain all the more special cases, but also be able to *deduce* them from the universal formula. […]. The ideal of a *scientific* concept here appears in opposition to the schematic general presentation which is expressed by a mere *word*. The genuine concept does not disregard the peculiarities and particularities which it holds under it, but seeks to show the *necessity* of the occurrence and connection of just these particularities."

the contemplated series.[144] As Cassirer often remarks,[145] physical theories of the modern period aimed foremost to transform sensible data into measurable data, the difficulty then focusing around the interpretation of the concept of *pure space*. Either space is conceived as a fixed frame within which matter is contained, in which case space remains prisoner of the imagination,[146] or it appears as a system of relations without substantial foundation, in which case it is unknown how it makes possible the objectivity of determinations.[147] This conception of space rests upon a strict separation between matter *in* space and the form *of* space which endures until the advent of Kant[148] and which derives from the clout of the mechanistic model instituted as a paradigm for all of physics during the seventeenth century. However, the progressive evolution of science after Newton and particularly the discovery of the concept of energy as described in the equations of thermodynamics developed by Maxwell [1831–1879], steers physics away from a mechanistic explanation, even if it remains possible after this date to interpret the concept of energy in the terms of the physics of force, as shown by Helmholtz [1821–1894].[149]

It nevertheless remains that it is already another conception of physics which predominates, that of the physics of "principles" which Cassirer opposes to a physics

---

[144] E. Cassirer, *Substance and Function*, pp. 22–23 (ECW 6, p. 22): "But precisely to the extent that the concept is freed of all thing-like being, its peculiar functional character is revealed. Fixed properties are replaced by universal rules that permit us to survey a total series of possible determinations at a single glance."

[145] E. Cassirer, *Einstein's Theory of Relativity*, p. 357 (ECW 10, p. 8): "What we possess in them are obviously not reproductions of simple things or sensations, but theoretical assumptions and constructions, which are intended to transform the merely sensible into something measurable, and thus into an "object of physics" that is, into an object for physics. Planck's neat formulation of the physical criterion of objectivity, that everything that can be measured exists, may appear completely sufficient from the standpoint of physics; from the standpoint of epistemology, it involves the problem of discovering the fundamental conditions of this measurability and of developing them in systematic completeness."

[146] This is the critique which Leibniz addresses to the Cartesian conception of substance; E. Cassirer, *The Philosophy of Symbolic Forms*, vol. 3, p. 456 (ECW 13, p. 529): "Whereas Descartes had criticized the Aristotelian explanation of nature for failing to recognize the limitations of sensation as such and to transcend them in principle, Leibniz attacks the Cartesian definition of substance for remaining wholly within the boundaries of that which can be represented intuitively; for this made the imagination into a judge over the understanding."

[147] Cf. supra § 134.

[148] E. Cassirer, *The Philosophy of Symbolic Forms*, vol. 3, p. 459 (ECW 13, p. 533): "The axiom that space itself and what fills it, what is substantial and real in it, are separate, that they may split conceptually into two sharply divided modes of being, is taken from the system of classical mechanics. But with this of course Kant's theory of pure intuition and the whole relation he sets up between the transcendental analytic and the transcendental aesthetic, runs into a difficulty which was bound to become apparent as soon as this axiom itself began to be questioned—as soon as the theory of classical mechanics gave way to the general theory of relativity."

[149] E. Cassirer, *The Philosophy of Symbolic Forms*, vol. 3, p. 461 (ECW 13, p. 535): "The task of physical science, in this view, is to reduce natural phenomena to invariable attractive and repellent forces, whose intensity depends on the distance between masses. If we start from this postulate and from Newton's general laws of motion, the principle of the conservation of energy seems essentially to reduce itself to the mechanical principle of the conservation of active forces."

of "models": it is no longer a matter of relating concepts to an underlying mechanical model which would be omnipresent and which in the end would come to replace substance in its foundational role, but rather of identifying the equations enabling to account for the *transformations* of physical concepts the ones into the others, without however identifying the ones with the others on a substantial level (for example, heat with motion and reciprocally).[150] The concepts of physics require no more but to first be conceived intuitively, but it is on the contrary at the end of the process of interpretation, inasmuch as it forms a coherent totality of propositions stemming from principles, that it becomes possible to return to a direct intuition of what is perceived. In this perspective, Cassirer opposes the physics of "models" of the seventeenth and eighteenth centuries—which he compares with "words", and which remains, in Kant's works, captive of a relation of resemblance between concepts and space under the aegis of the imagination—to the physics of "principles" of the nineteenth and twentieth centuries which he compares to "propositions"[151] which form a coherent set having no longer for interpretive framework relations of resemblance.[152] The relationship of the particular to the universal finds itself thereby completely modified because the sensible no longer plays the role of the particular with respect to a concept conceived as universal which would "apply" to it: the regulated translation of concepts thenceforth makes possible not only a physics of objects within pure space but also a physics of *planes of expression* between fields which no longer have the homogeneity of classical physics.[153] Quantification and measurement then appear as conceptual determinations depending upon a *system of symbols*[154] and not as deriving by similitude from isolated forms in nature, be they perceived directly or

---

[150]E. Cassirer, *The Philosophy of Symbolic Forms*, vol. 3, p. 462 (ECW 13, pp. 535–536): "Motion is transposed into gravitational force and gravitational force into motion, but we cannot infer that the two are identical; and the same is true for all spheres of phenomena which the law of the conservation of energy teaches us to connect by fixed numerical measurements, by definite equivalence."

[151]E. Cassirer, *The Philosophy of Symbolic Forms*, vol. 3, p. 461 (ECW 13, p. 535): "Thus to return to our comparison between physical and linguistic thinking, we might say that the progress from "model" to "principle" comprises n intellectual achievement similar to that of language when it advances from the word to the sentence: it is with the recognition of the pre-eminence of the principle over the model that physics begins, as it were, to speak in sentences rather than in words."

[152]E. Cassirer, *The Philosophy of Symbolic Forms*, vol. 3, p. 465 (ECW 13, p. 540): "The reality that we designate as a "field" is no longer a complex of physical things, but an expression for an aggregate of physical relations. When from these relations we single out certain elements, when we consider certain of its positions by themselves, it never means that we can actually separate them in intuition and disclose them as isolated intuitive structures. Each of these elements is conditioned by the whole to which it belongs; in fact it is first defined through this whole."

[153]E. Cassirer, *The Philosophy of Symbolic Forms*, vol. 3, pp. 461–462 (ECW 13, pp. 535–536): "[...] this law [the principle of the conservation of energy] signifies nothing other than a universal relation which links together the diverse spheres of physical phenomena, which makes them quantitatively comparable and commensurable."

[154]E. Cassirer, *The Problem of Knowledge...*; vol. 4, p. 114 (ECW 5, p. 133): "Here a particular symbol can never be set over against particular object and compared in respect to its similarity. All that is required is that the order of the symbols be arranged so as to express the order of the phenomena."

modelized in an abstract space.[155] By "system of symbols", one must read a certain mode of discourse which is not only mathematized—this was already the case during the seventeenth and eighteenth centuries—but which is *formalized*, in the sense that it does not depend on intuition but consists foremost in a set of principles to which a universal scope is attributed in a given reference system, for performing a global measurement of phenomena without reproducing them in isolation in the form of an image. Cassirer declares[156]:

The schematism of images has given way to the symbolism of principles.

This is what enables him to accomplish a number of epistemological analyzes pertaining to four fields: Maxwell's electrodynamics, Einstein's relativity, quantum mechanics and theoretical biology.

In order to explicate Cassirer's approach, we will only concentrate on two of these fields, relativity and theoretical biology, in a non-exhaustive manner. The reasons for this choice stem from the objective that Cassirer had set for himself, that is, the transcendental analysis of the extensions internal to the natural sciences after Kant. Such a starting point indeed requires, at least in the beginning, to return to a strict Kantian viewpoint and to the distinction it operates between two modes of judgment: the "determinant judgment" and "reflective judgment". A judgment is said to be "determinant" when it operates a synthesis in space-time from concepts provided by understanding; a judgment is said to be "reflective" when such synthesis is not performed due to a lack of concepts to schematize in space-time.[157] These two types of judgments may be characterized as they are deployed in the extension internal to the natural sciences after Kant in the following manner. In the first type, judgment would be "determinant" because it would be possible to integrally specify from outside the conditions of evolution of a physical system in which objects conceived to be of absolutely any type evolve: the instance of judgment would then be conceived as

---

[155]E. Cassirer, *Substance and Function*, p. 149 (ECW 6, p. 161): "The sensuous quality of a thing becomes a physical object, when it is transformed into a serial determination. The "thing" now changes from a sum of properties into a mathematical system of values, which are established with reference to some scale of comparison. Each of the different physical concepts defines such a scale, and thereby renders possible an increasingly intimate connection and arrangement of the elements of the given. [...]. In this logical connection, we first see the "objective" value in the transformation of the impression into the mathematical "symbol". It is sure that, in the symbolic designation, the particular property of the sensuous impression is lost; but all that distinguishes it as a *member of a system* is retained and brought out. The symbol possesses its adequate correlate in the *connection* according to law, that subsists between the individual members, and not in any constitutive part of the perception; yet it is this connection that gradually reveals itself to be the real kernel of the thought of empirical "reality"."

[156]E. Cassirer, *The Philosophy of Symbolic Forms*, vol. 3, p. 467 (ECW 13, p. 542).

[157]Kant, *Critique of Judgement*, Introduction, IV: "Judgement in general is the faculty of thinking the particular as contained under the Universal. If the universal (the rule, the principles, the law) be given, the Judgement which subsumes the particular under it (even if, as transcendental Judgement, it furnishes a priori the conditions in conformity with which subsumption under that universal is alone possible) is determinant. But if only the particular be given for which the universal has to be found, the Judgement is merely reflective." (*Critique of Judgement*, translated by J. H. Bernard, MacMilland & Co., London, 1914).

"outside of the world" and as capable of describing it integrally from the exterior. In the second type, judgment would be said to be "reflective" because short of being able to integrally determine from the outside a physical system's conditions of evolution (in particular, time would no longer constitute a parameter external to the system at hand, in the Newtonian sense), the knowing subject, engaged in the world, would be capable of recognizing in the form of objects the sufficient perceptual conditions for conceiving the relations of similitude and difference between them.

The two chosen examples—the theory of relativity and theoretical biology—thus aim to cover the field of judgment in its two modalities, that is, determinant judgment for relativity and reflective judgment for theoretical biology. We will see, at the end of the analysis, to what extent the Kantian distinction between these two types of judgment remains relevant or not in the eyes of Cassirer.

2.3 The Post-Kantian Extension of the Natural Sciences

2.3.1 The Case of Relativity

We know that before Einstein, the unification of Galileo's mechanics and of Maxwell's electro-dynamics proved impossible owing to the fact that Galileo's mechanics rested upon the universal principle of the relativity of motion whereas Maxwell's electro-dynamics rested on the discovery of the invariance of light's propagation speed regardless of motion in the physical system of reference. Contrarily to the firstly admitted viewpoint consisting in the supposition of an absolute "locus", ether,[158] serving as the invisible support for the invariance of the speed of light, Einstein's approach consisted, in 1905 within the context of special relativity, in extending the principle of the relativity of mechanical motion to electro-dynamics all the while affirming the invariance of the speed of light. Such an approach would be admissible only if the notions of the identity of measurement in time (simultaneity) and in space (distance) lost their status as absolute standards and became dependent on the reference system's state of motion.[159] However, the notion of "reference system" needed itself to be theoretically founded because it had only been observed empirically, be it, for example, in the case of Galileo's or Newton's fixed

---

[158] E. Cassirer, *Einstein's Theory of Relativity*, p. 376 (ECW 10, pp. 30–31): "Lorentz's assumption appeared above all to be epistemologically unsatisfactory because it ascribes to a physical object, the ether, definite effects, while at the same time it results from these effects that ether can never be an object of possible observation." It is for the same reason that Cassirer dismisses, with respect to the epistemology of quantum mechanics, the "hidden variable" theories because they derogate from the Leibnizian principle of observability.

[159] E. Cassirer, *Einstein's Theory of Relativity*, pp. 371–372 (ECW 10, p. 25): "The decisive step is taken when it is seen that the measurements, to be gained within a system by definite physical methods of measurement, by the application of fixed measuring-rods and clocks, have no "absolute" meaning fixed once for all, but that they are dependent on the state of motion of the system and must necessarily result differently according to the latter."

stars or in the case of light in electro-dynamic theory or even in special relativity. From this perspective, the theory of general relativity achieved two major results. First, it extended *a priori* the notion of relativity to the idea of *any* reference system, by showing that the notion of reference system, first defined within the framework of a Euclidean system of coordinates, could be generalized and become absolutely generic if recourse was made to non-Euclidean coordinate systems. On the other hand, it managed to determine *a priori* the relations between reference systems by means of mathematically regulated permutations of coordinate systems in which the chosen reference systems made measurements possible.[160] Thus, physical phenomena considered as *different* from the standpoint of coordinate systems enabling to perform their measurement nevertheless received mathematical expressions of the *same* form. Then, what had remained unexplained up to that point in the framework of Newtonian mechanics, that is, the reason for which the universal attraction of bodies is proportional to their inert mass, received at least a formal translation since it was possible to show the *functional* equivalence between inertial motions describable by the classical means of Newtonian dynamics and the motions submitted to a gravitational field describable by geometrical means, once it had been generalized towards non-Euclidean geometries.[161]

The theory of relativity has the particularity, in what concerns the idea of the extension of the functional viewpoint in physical theory, of possessing two characteristic features which make it exemplary: on the one hand, it brings to its peak the movement of modern physics initiated by Galileo and going so far as to make space and time functional themselves rather than only the notion of motion with respect to a given reference point; on the other hand, it doubly goes beyond the letter of the Kantian epistemological framework for which it is the *pure intuition* of space and time which makes measurement possible by conceiving this intuition as being *Euclidean* in nature.

Concerning the first point, it must first of all be noted that for Cassirer, a purely functional perspective concerning space and time is indeed to be credited to the theory of relativity, but that it is not that which constitutes its specific philosophical originality because it had already been virtually contained in the adoption by Galileo

---

[160]E. Cassirer, *Einstein's Theory of Relativity*, p. 406 (ECW 10, pp. 66–67): "In place of the rigid rod which is assumed to retain the same unchanging length for all times and places and under all particular conditions of measurement there now appear the curved coordinates of Gauss. [...] measurements in general different from each other result for each place in the space- time continuum."

[161]E. Cassirer, *Einstein's Theory of Relativity*, pp. 401–402 (ECW 10, p. 61): "What was previously done in the Newtonian theory of gravitation by the dynamics of forces is done by pure kinematics in Einstein's theory, *i.e.*, by the consideration of different systems of reference moving relatively to each other."

of a principle of relativity of inertial movement.[162] There is therefore here a full continuity between physical theory since the beginning of the modern era and the theory of relativity: the latter integrates easily into the Kantian epistemological apparatus which conferred no independent reality to space and to time,[163] but which already made them into a functional unit.

In what concerns the overstepping of the letter of the Kantian epistemological framework, two points, for Cassirer, need to be emphasized. First, a transformation must affect the notion of *pure intuition* in the new framework imposed by the theory of relativity. The notion of pure intuition requires, according to the Kantian formulation, an act of synthesis of diversity which manifests as a homogenous and integral order of coexistence and of succession.[164] Due to this, the disappearance of the invariance of the distance between two points as well as of the simultaneity of events undermines the very idea of a pure intuition capable of uniting into an intuitive whole the diversity of loci and events. But, as emphasized by Cassirer, even if all metric determinations corresponding to geometric forms thenceforward occur within a *heterogeneous* space-time, the *necessity* for a synthesis of spatio-temporal diversity remains[165]: only, this synthesis of the diverse no longer operates within what Kant called pure intuition, but indeed at the level of what he called *understanding*, which poses the rules of permutation between the systems of reference generating various metrics.[166] A corollary immediately stems from this directly philosophical point

---

[162] E. Cassirer, *Einstein's Theory of Relativity*, p. 357 (ECW 10, p. 8): "That physical objectivity is denied to space and time by this theory must, as is now seen, mean something else and something deeper than the knowledge that the two are not things in the sense of "naive realism". For things of this sort, we must have left behind us at the threshold of exact scientific physics, in the formulation of its first judgments and propositions."

[163] E. Cassirer, *Einstein's Theory of Relativity*, p. 412 (ECW 10, p. 73): "Accordingly, when Einstein characterizes as a fundamental feature of the theory of relativity that it takes from space and time "the last remainder of physical objectivity" it is clear that the theory only accomplishes the most definite application and carrying through of the standpoint of critical idealism within empirical science itself."

[164] E. Cassirer, *Einstein's Theory of Relativity*, p. 417 (ECW 10, pp. 79–80): "In fact, the point at which the general theory of relativity must implicitly recognize the methodic presupposition, which Kant calls "pure intuition" can be pointed out exactly. It lies, in fact, in the concept of "coincidence" to which the general theory of relativity ultimately reduces the content and the form of all laws of nature. If we characterize events by their space-time coordinates $x_1, x_2, x_3, x_4, x'_1, x'_2, x'_3, x'_4$, etc., then, as it emphasizes, everything that physics can teach us of the "essence" of natural processes consists merely in assertions concerning the coincidences or meetings of such points. We reach the construction of physical time and of physical space merely in this way; for the whole of the space-time manifold is nothing else than the whole of such coordinations."

[165] E. Cassirer, *Einstein's Theory of Relativity*, p. 418 (ECW 10, p. 80): "Not that the theory, as has been occasionally objected, presupposes space and time as something already given, for it must be declared free of this epistemological circle, but in the sense that it cannot lack the form and function of spatiality and temporality in general."

[166] E. Cassirer, *Einstein's Theory of Relativity*, pp. 439–440 (ECW 10, p. 104): "The step beyond him [Kant], that we have now to make on the basis of the results of the general theory of relativity, consists in the insight that geometrical axioms and laws of other than Euclidean form can enter into this determination of the understanding, in which the empirical and physical world arises for us, and that the admission of such axioms not only does not destroy the unity of the world, i.e.,

of view: the necessity for having recourse to non-Euclidean geometries in order to establish the full coherence of the notion of the relativity of reference systems. This point indeed constitutes a *philosophical* revolution which Cassirer was undoubtedly one of the first to notice, in his case as early as in 1910,[167] because this perspective not only breaks with the Kantian theory of the intuition of space and of time founded upon Euclidean geometry and supposed, up till then in Cassirer's view, to establish the epistemological framework for what makes measurement possible in physics,[168] but still furthermore, it promotes *generalized functionalization in knowledge in general*, that is, the discovery of the fact that the act of knowledge itself possess its *own* conditions of validity, as Euler had remarked concerning the particular case of Newtonian physics. What remains then is the philosophical difficulty of determining *with respect to which invariant* these systems of reference depending upon specific acts of judgment can be interpreted as *variable*.

For Cassirer, the answer to this question consists in remarking that *only the expressive dimension* such as it manifests first and foremost but not only in language gives an idea of the articulation between invariant and variation.[169] In the particular case of

---

the unity of our experiential concept of a total order of phenomena, but first truly grounds it from a new angle, since in this way the particular laws of nature, with which we have to calculate in space-time determination, are ultimately brought to the unity of a supreme principle, that of the universal postulate of relativity."

[167]Indeed, in E. Cassirer, *Substance and Function*, p. 111 (ECW 6, p. 119): "That experience in its present scientific form gives no occasion to go beyond Euclidean space is expressly admitted even by the most radical empiristic critics. From the standpoint of our present knowledge, they also conclude, we are justified in the judgement that physical space "is to be regarded as positively Euclidean". Only we must not exclude the possibility that in a distant future perhaps changes will take place here also. If any firmly established observations appear, which disagree with our previous theoretical system of nature, and which cannot be brought into harmony with it even by far-reaching changes in the physical foundations of the system, then, all conceptual changes within the narrower circle having been tried in vain, the query may arise whether the lost unity is not to be reestablished by a change in the "form of space" itself." The theory of general relativity which directly solicits the notion of non-Euclidean geometry had been elaborated by Einstein between 1905 and 1915 and Cassirer had probably not yet fully assimilated it. But he was nevertheless able, as early as 1910, to *philosophically conceive* of the adoption of a non-Euclidean point of view even if it was *not yet* relevant in physics.

[168]E. Cassirer, *The Problem of Knowledge...*; vol. 4, p. 51 (ECW 5, p. 58): "It is characteristic of the thought of modern geometry, as of modern physics, that in both cases the process of measuring is recognized more and more clearly to be a *problem*, a logical and epistemological one. What had been constantly practiced before in the two sciences was seen from a different angle as soon as one tried to make wholly clear the assumptions underlying mensuration. In geometry, the discovery of the non-Euclidean systems first demanded such a view and opened the way to it. In so doing, furthermore, the discourse, far from raising a question about the a priori character of geometry, actually brought it into greater prominence. The concept of space as such, when taken in its most general sense as "order of coexistence", appeared to be an underivable basic concept. The problem of the significance and role of experience arose only when one dealt with the specification of this general space concept by introducing a measurement."

[169]E. Cassirer, "The Concept of Symbolic Form in the Construction of the Human Sciences", in *E. Cassirer, The Warburg Years (1919–1933)*, p. 75 (ECW 16, p. 78): "It is a question of taking symbolic expression, that is, the expression of something "spiritual" through sensory "signs" and "images", in its most general signification; it is a question of asking wether this form of expression,

the linguistic expression, the possibility of a specialization of language with respect to the objectives a particular activity assigns for itself, nevertheless makes it possible to conceive of its differentiated unity and its fundamental mode of being consisting in continuously rearticulating the material and conceptual aspects of signs in expressive forms. The most direct consequence then consists in remarking that, if it is put into correspondence with its symbolic dimension, the functional point of view is no longer restricted to the natural sciences, as Euler and Kant may still have believed, but that it also informs the language sciences—as witnessed, according to Cassirer, by the history of knowledge since the Renaissance—*and beyond*; not only the act of *knowledge*, but human *activities* in general as they all develop as modalities of expression.

2.3.2 The Case of Theoretical Biology

As Cassirer often stresses, the question of a transcendental legality specific to objects called "living" had been treated by Kant from the standpoint of the possible modes of knowledge and not from the metaphysical standpoint of an ontology.[170] In this perspective, the possibility for "biology" did not distinguish itself, in the view of a transcendental analysis, from the possibility for "physics". The problem however did not pose itself in the exact same terms in both cases. In the case of physics, the post-Kantian developments had required to go beyond the framework defined by the *Critique of Pure Reason* since it had been necessary to renounce keeping with a theory of schematism in order to justify the advances in physics. In the case of biology, it was rather necessary to *return* to the Kantian viewpoint such as it had been exposed in the antinomy of the faculty of judgment of the *Critique of Judgment* in order to account for the contemporary advances in theoretical biology during the age of Cassirer.

Indeed, over the course of the nineteenth century and in the beginning of the twentieth century, it was the quarrel between mechanism and vitalism which seemed to have caught most of the attention in the philosophy of biology, this revealing the legacy of the theoretical framework established by Kant. In hindsight, this quarrel seems essentially to have corroborated, for Cassirer, the validity of the Kantian point of view according to which the antinomy between mechanism and finality would dissolve by itself when one refused for either position to take a dogmatic stance and when one adopted a critical stance towards them. Yet Cassirer did not justify the critical point of view in question using Kantian arguments, in particular because he rejected the difference between "determinant" judgment and "reflective" judgment,

---

with all of its different possible applications, is grounded by a principle that marks it as a closed and unified fundamental process. Thus, what should not be asked here is what the symbol signifies and achieves in any *particular* sphere, what it signifies or achieves in art, myth, or language, but how far language as a *whole*, myth as a *whole*, art as a *whole*, carry within them the general character of symbolic configuration."

[170] E. Cassirer, *The Problem of Knowledge...*; vol. 4, p. 120 (ECW 5, p. 140): "Kant did not proceed from the existence of "things in themselves" but only analyzed the knowing of them."

which he then considered to be obsolete.[171] It is indeed rather an argument that is *historical without being empirical* that he advanced, that is, the extension of physics beyond the Newtonian framework,[172] and in particular the appearance of relativistic physics as he had described himself: since physics had itself progressively steered away from the mechanistic model of forces in order to adopt a point of view which Cassirer calls the physics of "principles", there was no reason, in the epistemology of biology, to keep with the sterile opposition between mechanism which, by limiting itself to a narrow reductionism, seeks to translate into mechanistic terms the central concept of biology which is finality and, on the other hand, vitalism which, by adding to mechanical forces other forces said to be "vital", purports to account for that which produces organization in matter conceived to be inanimate.

For Cassirer, one of the principles governing research in the natural sciences of the twentieth century is the concept of *totality*, be it in physics with the concept of field or in biology with the concepts of regulation and of regeneration.[173] It is this concept of totality which enables the conflict between mechanism and vitalism to evolve.[174] For mechanical philosophy, renouncing a dogmatic position consists in renouncing the methodological option of reductionism which is no longer relevant due to the transformations in physics: the latter operates by synthesis of already existing theories within a new, more *encompassing* theory, and not by the reduction of an existing theory to another existing theory that is deemed to be more fundamental, as has clearly been established by relativistic physics. For its part, vitalism must renounce the overly anthropomorphic concept of finality and keep with the operational concept of totality which must no longer be conceived as a simple "as if", in the manner of Kant, but indeed as possessing a *reality*. The real character attributed to the organic totality does not however signify an occult return to ontology as it was conceived by pre-Kantian metaphysics. Indeed, totality appears, in the context of biology, as essentially *diversified*, inasmuch as it is always circumscribed within the environment

---

[171] E. Cassirer, *The Problem of Knowledge*...; vol. 4, p. 211 (ECW 5, p. 244): "We need not follow out this discrimination made by Kant, since it depended upon the form of the problem in his own day, and the problem has undergone a fundamental change since then both in physics and in biology."

[172] E. Cassirer, *The Problem of Knowledge*...; vol. 4, p. 211 (ECW 5, p. 245): "For modern physics has gradually freed itself of the chains imposed on it by the mechanistic view and developed a new ideal of science."

[173] E. Cassirer, *The Problem of Knowledge*...; vol. 4, p. 213 (ECW 5, p. 246): "In psychology, similarly, it demands the transition from a psychology of separate elements to Gestalt psychology."

[174] It should be noted that for Cassirer, Darwinism appears as a particular variant of *critical teleology*, quite removed from the vulgate which would make it into a theory rigorously excluding any reliance upon finality; E. Cassirer, *The Problem of Knowledge*...; vol. 4, p. 166 (ECW 5, p. 192): "Here the concepts of purpose have their secure place; they prove themselves to be not only admissible but absolutely indispensable. Not only the answer that Darwinism yields but even its formulation of the problem is indissolubly bound up with the purposive idea. The concepts of 'fitness', 'selection', 'struggle for existence', 'survival of the fittest'—these all have plainly a purposive character and they exhibit a theoretical organization quite different from that of the concepts of the mathematical sciences. The constant use that Darwinism makes, and must make, of these concepts would alone suffice to show that in opposing a definite form of metaphysical teleology it in no way renounces "critical teleology"."

specific to each species of organisms. Cassirer demonstrated in this respect that the works by von Uexküll converge with the Goethean idea according to which each species lives in its own world corresponding to the potentialities of its anatomical structure.[175] The shift from the concept of finality to that of totality is therefore also the shift from the regulating Kantian idea of finality to a Goethean realism of the form justified by the fact that it is thenceforth a matter of adopting a point of view according to which universality is not given at the onset, but to be constructed by a structural extension, as was the case with the project defended by Klein in geometry.

This realism of totalities requires transforming the relation between the particular and the universal and this is incidentally the reason why the Kantian distinction between determinant judgment and reflective judgment has lost validity: as Goethe had already remarked, *the single occurrence may concentrate within itself more than one case*.[176] The summation of these individual cases makes possible the logical relation of subsumption but it no longer constitutes the relevant concept for thinking about the relation between the particular and the universal. This would require, from this new standpoint, to be replaced by the notion of *pregnancy* which concentrates, in a nevertheless unique occurrence, the *exemplarity* of the cases. What replaces, in pregnancy, the simple one-by-one summation of elements is thus a relation founded on the recognition, by the knowing subject, of exemplarity. But whereas the universal was conceived as an ideality as opposed to the reality of the individual cases in the relation of subsumption, *exemplarity has for realistic feature that it can embody the universal in a nevertheless unique case*. Thus, the exemplary case no longer integrates the logical relation of subsumption and it is in this sense that it is of a *symbolic* nature, but in a sense which is no longer the one put forth in the *Critique of Judgment*.[177] Thus, the notion of organic totality, insofar as it is *real*, integrates

---

[175] E. Cassirer, *The Problem of Knowledge…*; vol. 4, p. 205 (ECW 5, p. 237) after a long quotation of Goethe's poem "Die Metamorphose der Tiere" (The Metamorphosis of Animals), Cassirer claims: "It is remarkable to see how exactly the plan and development of Uexküll's biology conformed in every particular with this view of Goethe." The relations between Cassirer and von Uexküll are complex: both belonging to the Kantian legacy and having both insisted on the capital role played by semiotic mediations in the case of the human cultural world, Cassirer develops, starting with the ethological works of von Uexküll, an *optimistic* view of mediation whereas the latter developed, based on the same works, a pessimistic view regarding any form of mediation which will lead him to favor for the human world a political order replicating the immutable order of animal societies. Cf. F. Stjernfelt, "Simple Animals and Complex Biology. The double von Uexküll inspiration in Cassirer's philosophy", in *Synthese*, Vol. 179, No. 1, 169–186, 2009.

[176] E. Cassirer, „Goethe und die mathematische Physik. Eine erkenntnistheoretische Betrachtung", ECW 9, p. 277: "In the battle that Bacon and Galileo would wage against each other concerning the concept of experience, he [Goethe] would stand, determined and confident, on the side of the latter. For he would see in him the tangible proof that the Universal neither results from the mere summation of isolated cases nor consists in a mere aggregate of particulars but that one case is worth a thousand for the genius researcher in physics in so far as, in his mind, he overlooks the conditions of the exceptional case, so that through a continuous variation of these conditions he can bring to mind the whole range of possible cases and grasp them in their determined law."

[177] E. Cassirer, *The Problem of Knowledge…*; vol. 4, p. 146 (ECW 5, pp. 168–169): "The relation between the two [the particular and the universal] according to Goethe, is not one of logical subsomption but of ideal or "symbolic" representation. The particular represents the universal, "not

from the onset a *symbolical* dimension—if this is intended to mean the intensity of a pregnancy.

2.3.3 Modern Science and the Transcendental Role of the Symbolic Process

Now that the above examples have made clearer what Cassirer meant by the extension of the transcendental legality beyond the historical state in which the natural sciences found themselves during the age of Kant, it would now be suitable to close the inquiry by trying to see if the term "symbolic" has evolved and if it has not itself acquired new determinations over the course of the analysis.

At this stage, it should first be noted that the *semiotic* question of symbolism introduces itself in the stead of a theory of schematism which remains prisoner of a problematic relation of adequation between concepts and images inasmuch as the rule, as a synthetic production, must have *corresponding* particular instances which are intuited objects.[178] Cassirer expressly compares the usage he makes of the notion of symbolism when he describes the natural sciences of the nineteenth and twentieth centuries with its usage by Hilbert in mathematics,[179] but he immediately shows

---

as dream and shadow, but as momentarily living manifestation of the inscrutable." (J. W. Goethe, *Maximen und Reflexionen*, n°314, p. 59). For the true naturalist one "pregnant instance", not countless, scattered observations, exhibits the "immanent law" of nature. For this relationship neither "deduction" nor "induction", as there are ordinarily called, is a pertinent and adequate expression."

[178]E. Cassirer, *The Philosophy of Symbolic Forms*, vol. 3, p. 464 (ECW 13, p. 538): "The physics of the nineteenth century and the early twentieth century could not have achieved this advance to principles of ever increasing breadth and universality, could not have attained its present intellectual heights, if it had not steadily freed itself from the barriers not only of sensation but also of intuition and geometrical-mechanical "representation"."

[179]In this regard, Cassirer cites Weyl's book describing Hilbert's position; E. Cassirer, *The Philosophy of Symbolic Forms*, vol. 3, pp. 467–468 (ECW 13, p. 543): ""Intuitive space and time", writes Weyl, "may not serve as the medium in which the physics of the external world is constructed but are replaced by a four-dimensional continuum in the abstract arithmetical sense. [...] Thus what remains is ultimately a symbolic construction in the exact sense as that carried out by Hilbert in mathematics." (H. Weyl, *Philosophie der Mathematik und Naturwissenschaft*, 1927, p. 80)." Interpreting Hilbert's theory of the symbol would require in itself some development, especially given that Cassirer bases himself on Weyl's authority to place Hilbert under his own banner, whereas there is a bias in wanting to present Hilbert's point of view as that which would endow the "symbolic" approach with a role, in the Cassirerean sense, to account for the continuity of real numbers from a logical standpoint. It should indeed be noticed that the formalist point of view in demonstration theory as it was developed by Hilbert rests on a finitary approach derived from the arithmetics of natural integers, whereas it is here a matter of a quadridimensional *continuum* which would only be reducible to the finitary approach if the Cantorian issue of the cardinality of continuity had indeed found a solution in finitary terms—this having remained an open question during the age of Hilbert. What Cassirer sought here is probably rather the role played for Hilbert by the notion of "implicit definition": It is not the isolated axiom, but rather the *groups of axioms* which, once they have been expressed linguistically, determine, by the reciprocal relations they induce, a configuration of meaning for the concepts which are contained within them. The content of concepts in terms of meaning thus depends on the way in which the axioms have been grouped, this being relatable to

that the formalist interpretation of the symbol as it had been elaborated by Hilbert in his theory of demonstration does not however enable to philosophically justify the transcendental character of mathematics.[180] Indeed, for Cassirer, the attention that Hilbert gives to the judicatory virtues of formalism enable, already in his works on the foundations of geometry, to essentially justify *afterwards* the conformity of the derivation modes of propositions in terms of formal consistency, that is, of non-contradiction, from the fundamental axioms of mathematics. But formalism does not enable to account for the reasons which place mathematics at the very center of the possibility for "natural sciences". *Now, it is indeed this transcendental character which Cassirer tries to attribute to symbolism.* On what does it rest? To answer this question, it is first necessary to briefly summarize the five characteristics which Cassirer uses over the course of his epistemological progression to describe what he means by "symbolic".

Firstly, Cassirer will have at times used the adjective "symbolic" in a pejorative manner. In such cases, he would qualify as "symbolic" that which is only *allegorical*[181]: when phenomena or events are conceived in isolation and that they are affixed with numeric or geometric signs meant to ensure their objective determination, they are only actually given but a simple *name* and the possibility of determining what constitutes the link between the phenomena themselves, in particular their causal relations, is lost.[182]

The positive usage of the adjective "symbolic" is linked with the advent of modern science starting in the seventeenth century and it possesses four distinctive features.

It firstly designates the *mathematical space* in which are studied the relations between idealized objects submitted to laws. This is the meaning of the adjective "symbolic" which Cassirer associates with the works of Galileo, Descartes, and Kepler.[183]

He then designates *arrangement in series* which, by ensuring the order and connection between signs in the process of knowledge, makes certitude possible. This characteristic was brought to light by Leibniz, particularly when he described the purely algebraic moment of "blind thought" which fully rests upon processualities

---

Cassirer's idea of symbols not having meaning when taken in isolation but rather inducing meaning when grouped.

[180]E. Cassirer, *The Philosophy of Symbolic Forms*, vol. 3, p. 387 (ECW 13, p. 446): "But even if his [Hilbert's] theory of proof should fully achieve this aim, the logician and epistemologist may be permitted to ask whether the forces here enlisted for the protection of the mathematical state power are the same as those which have established and never ceased to extend the preeminent position of mathematics in the realm of the intellect. Formalism is an incomparable instrument for the discipline of mathematical reason, but in itself it cannot explain the content of mathematics, or justify its transcendental use."

[181]cf. § 12.

[182]Cf. § 114.

[183]Cf. § 12.

inherent to signs, temporarily cut off from any relation to their referent.[184] It is to this distinctive feature that the fecundity of modern science must be attributed.[185]

Cassirer also uses the adjective "symbolic" when he seeks to describe the notion of form as a *deployment through time of an organized structure*, in line with the works of Goethe on morphology.[186] This acceptation of the adjective "symbolic" then breaks with the Kantian distinction between determinant judgment and reflective judgment which were opposed as two methods of apprehension of the relation between the universal and the particular, the first type of judgment consisting in an *a priori* relation of logical subsumption of the particular under the universal, and the second type consisting in an only "symbolic" attempt—that is, one which is not given *a priori*—of presentation of the universal based on the particular. As we have just seen, the "Goethean" acceptation of the adjective "symbolic" described here indeed goes beyond the purely logical distinction between the particular and the universal: a unique case, although it is "particular" from a logical standpoint, can nevertheless concentrate within itself more than what a sum of cases represents. The logical notion of subsumption thus appears as uselessly restricting to summation alone what symbolic *exemplarity* can in itself directly manifest. This way of understanding the adjective "symbolic" pertains to aesthetics, taken in the broad sense of a theory of the organization and diffusion of forms, and which is not directly mathematical in the sense that in its principle it makes no recourse to a geometrical space nor to an algebraic process of serialization but only to the idea of the perpetuation of an internal coherence and this, starting at the moment of perception.[187] It is precisely the possibility of a *mathematical* determination of the perpetuation of the internal coherence of structures that the last stroke will concern.

Cassirer attempts indeed, in a last acceptation of the adjective "symbolic", to find a way to unite the Goethean sense of structural nature with the mathematical sense—both geometrical and algebraic—by discarding the Kantian opposition, as it is described, between determinant judgment and reflective judgment. According to the Kantian acceptation, any presentation of a concept in intuition is done either in a schematic manner by means of demonstration or in a symbolic manner by analogy

---

[184]Cf. § 133.

[185]E. Cassirer, "The Influence of Language upon the Development of Scientific Thought", *The Journal of Philosophy*, vol. 39, n°12 [309–327], p. 320 (ECW 24, p. 127): "Kant has described this task by saying that the scope of a scientific theory is: Erscheinungen zu buchstabieren, um sie als Erfahrungen lesen zu konnen ("to spell phenomena in order to be able to read them as experiences"). Modern evolution of physics has shown us that science in this spelling of phenomena may follow different ways. It is not restricted to a special type of spelling and to a single alphabet; it is at liberty to choose various sets of symbols. But of course we can not use these symbols at random. We must find certain rules that determine their mutual relation and connection."

[186]Cf. § 21.

[187]E. Cassirer, *The Philosophy of Symbolic Forms*, vol. 3, p. 193 (ECW 13, p. 220): "As long as this view prevailed, the first presuppositions for any true phenomenology of perception was lacking. By restricting themselves in principle to the data of sensation, sensationalism and positivism had blinded themselves, as it were, not only to the symbol, but also to perception itself. For they had eliminated precisely the characteristic factor by which perception differs from mere sensation and grows beyond it."

in the case where the schema would be lacking,[188] the presentation of the concept in intuition escaping the mathematized register of objectivity. Then, a presentation of the concept in intuition is "schematic" when it pertains to determinant judgment and it is "symbolic" when it pertains to reflective judgment. Now, the shift operated by the classification of geometries by Klein based on the concept of group of transformation[189] as well as the possibility of making a choice among several geometries by integrating non-Euclidean geometries at the very core of physics starting with Einstein[190] have shown that *this "lacking" was precisely the way in which the quest for universality manifested itself in the sciences* because pregnant forms would therein manifest in an exemplary manner, in the Goethean sense. There is therefore a process at work at the very core of the advancement of sciences which Cassirer relates both to the *concept* and to *organization* without, however, relating it to the *schema*.[191] It is the *organizing value of concepts serving as principles of objectification* which Cassirer then describes as "symbolic".

The process at work in the sciences then requires further reflection regarding the nature of *objectification* which must also be re-conceptualized so that it may be neither identified with schematism (which was the case from Kepler to Kant) nor with the concept (which was the case from Leibniz to Hilbert), but first and foremost as *sense*. This locus of sense is precisely that of the *symbolic forms* which, by organizing the diversity of objectification modes, thenceforth perform the transcendental function previously related by Kant to schematism.

We have seen that the analysis of the modalities of objectification such as they appear in the natural sciences was linked with a true semiotic revolution which occurred during the Renaissance and which Cassirer began describing in 1906.[192] Afterwards, the analysis conducted by Cassirer regarding the progressive dissemination of the functional perspective within the natural sciences, a dissemination of which we have just roughly outlined the stages, nevertheless had an unexpected result regarding the semiotic assumptions of the method employed by Cassirer, that is, the full recognition of the fact that the analysis of the modalities of objectification,

---

[188] Kant, *Critique of Judgement*, § 59: "All intuitions, which we supply to concepts a priori, are therefore either schemata or symbols, of which the former contain direct, the latter indirect, presentations of the concept. The former do this demonstratively; the latter by means of an analogy (for which we avail ourselves even of empirical intuitions) in which the Judgement exercises a double function; first applying the concept to the object of a sensible intuition, and then applying the mere rule of the reflection made upon that intuition to a quite different object of which the first is only a symbol. Thus a monarchical state is represented by a living body, if it is governed by national laws, and by a mere machine (like a hand-mill) if governed by an individual absolute will; but in both case only symbolically."

[189] Cf. Chapter 1, § 23.

[190] Cf. Chapter 1, Conclusion.

[191] Cf. § 222. We can see that what distinguishes Cassirer here from his masters at Marburg is that *they lacked the concept of group* of *transformation* and, more broadly, the concepts which would have enabled to scientifically justify what was only the epistemological conviction of the well-foundedness of the broadening of the transcendental perspective to the internal *and external* developments of science.

[192] Cf. § 11.

given that they were at first entrenched within the natural sciences, needed to extend *beyond*. This broadening is not self-evident and it appears, in the least at first glance, to be problematic.[193] On the one hand, indeed, if we keep with the epistemology that is *internal* to the natural sciences as it is analyzed by Cassirer, we will notice that he insists on the fact that it is the search for greater *unity* which guides the evolution of the natural sciences towards ever-increasingly encompassing theories of reality—in particular in the case of relativity.[194] On the other hand, the extension of the issue of the modalities of objectification *beyond* the sole natural sciences stems, in Cassirer's view, from the observation that it has become *illusory* to consider reality in its unity and its simplicity.[195] How then may it be conceived that what makes possible the ever-increasing unity *within* the natural sciences is considered to be an illusion which should be abandoned *outside* of these sciences?

In order to understand this delicate point, it is necessary to return to the way in which Cassirer conceives of the quest for the universal since his analysis of Klein's works in geometry[196]: tracing limits around well-circumscribed sets of objects or events ensure their own specific individuation within a particular form and it is this differentiation of forms which makes possible the search for ever-increasingly universal or limitless invariants. This open and limitless search involves the exemplarity of certain forms which short-circuit the simple logic of the summation of particular cases, as found by Cassirer in the works of Goethe. In Cassirer's view, the *exemplary* role, in the Goethean sense, played by Einstein becomes clear: *as Einstein allows to escape the exclusively Euclidean frame of reference in physics, Cassirer allows to escape the exclusively physical frame of reference in epistemology.* Thenceforth, it becomes possible to speak of the *various forms of objectification* which are no

---

[193]It is usually only considered as a simple fact which appears at a certain moment of Cassirer's intellectual journey. Cf. D. Gawronsky, "Ernst Cassirer: His Life and Work", in *The Library of Living Philosophers* vol. 6, P. A. Schilpp ed., Northwestern University, 1949, p. 25sq.

[194]E. Cassirer, *Einstein's Theory of Relativity*, pp. 354–355 (ECW 10, p. 5): "And even the special theory of relativity is such that its advantage over other explanations, such as Lorentz's hypothesis of contraction, is based not so much on its empirical material as on its pure logical form, not so much on its physical as on its general systematic value. In this connection the comparison holds which Planck has drawn between the theory of relativity and the Copernican cosmological reform. The Copernican view could point, when it appeared, to no single new "fact" by which it was absolutely demanded to the exclusion of all earlier astronomical explanations, but its value and real cogency lay in the fundamental and systematic clarity, which it spread over the whole of the knowledge of nature."

[195]E. Cassirer, *Einstein's Theory of Relativity*, p. 446 (ECW 10, p. 112): "The postulate of relativity may be the purest, most universal and sharpest expression of the physical concept of objectivity, but this concept of the physical object does not coincide, from the standpoint of the general criticism of knowledge, with reality absolutely. The progress of epistemological analysis is shown in that the assumption of the simplicity and oneness of the concepts of reality is recognized more and more as an illusion. Each of the original directions of knowledge, each interpretation, which it makes of phenomena to combine them into the unity of a theoretical connection or into a definite unity of meaning, involves a special understanding and formulation of the concept of reality."

[196]Cf. Chapter 1, Conclusion.

longer entrenched in the sole analysis of their modalities in the natural sciences. It is therefore necessary to extend the epistemological field to *all* forms of objectification for which it is possible to analyze the modalities of individuation.

This represents a considerable change in perspective which will, starting with his 1920 book on relativity, thoroughly redefine Cassirer's ulterior work. This change in perspective very deeply affects traditional epistemological notions, particularly two among them: the difference between the particular and the universal on the one hand and the relation between the empirical and the historical on the other.

The difference between the universal and the particular, inherited from classical logic and adopted without modification by the transcendental perspective, had been completely transformed by Goethe. Cassirer had the insight of remembering this when the time came to describe the internal dynamics of the forms of objectification, in particular in the case of relativity.

In the same way, the relation between the necessary and the empirical was also transformed since the internal logic of the development of science presents a series of facts of which the necessity is however dependent on a history which is no more contingent than the relation of kinship between organisms of a same species: such is the case, for example, with the progressive dissemination of the functional point of view throughout the natural sciences. It is this double transformation of the until then fundamental epistemological notions which enables to give epistemological substance to the notion of *culture* as well as to the forever open set of forms of objectification. Neither necessary, nor contingent, nor universal at the onset, nor only particular, the notion of culture requires to redefine the individual or collective expressions of human subjects in a whole different manner than by using the above-mentioned classical opposition pairs. From this point of view, it is absolutely not a matter, for Cassirer, of devaluating the eminent role of the epistemology of the natural sciences or to keep with a relativistic and skeptical interpretation once the place they occupy in culture has been redefined. It is, however, necessary to give it a *new status* from the moment where it maintains relationships with other forms of objectification of which the full legitimacy is recognized. This legitimacy is no longer justified by a simple call to the transcendental Subject, but by a call to *meaning*: it is because they manifest meaning that the forms of objectification are both intrinsically diverse all the while preserving their kinship.

How must the notion of "form" be understood in this context? One would imme-diately think of the notion of the figure in space as a result of a process of delimitation with respect to a background, a process which makes possible the expressivity spe-cific to the figure. But it is only a matter of the most simple aspect of an otherwise more complex semiotic process, that of the *institution of signs*. Cassirer, when refer-ring to Humboldt's point of view regarding the particular case of language, indeed remarks that, by "internal form" of a language, Humboldt means the work specific to each language consisting in organizing sound material for expressive ends.[197] Cassirer then seeks to give to the notion of "internal form" a general philosophical scope by means of his reflection on symbols: the notion of "internal form" may not only characterize each particular language with respect to other languages, but language in general when it is compared with other forms of objectification. This means that any form of objectification has its own "internal form", which Cassirer calls its "symbolic form". Thus, it is thenceforth a matter of studying from a general philosophical point of view the forms of objectification as they have progressively been deposited in culture—in the almost chemical sense of the term—their dissemination, their endurance, and their sometimes hidden links, as Cassirer did when he studied the dissemination of the functional point of view throughout the modern history of the natural sciences by starting with the philological revolution of the Renaissance.

---

[197] E. Cassirer, *The Philosophy of Symbolic Forms*, vol. 1, p. (ECW 11, p. 103): "What we call the essence and form of a language is consequently nothing other than the enduring, uniform element which we cannot demonstrate not in any one phenomenon, but in the endeavor of the spirit to raise the articulated sound to the level of an expression of thought." [Humboldt, „Einleiting zum Kawi-Werk", *Werke*, 7, N°1, 46 ff]."

Part II
Semiotics

Chapter 3
The Semiotic Situation of Cassirer

Abstract This chapter endeavors to examine the relationship between science, myth and their common semiotic foundation which has two aspects. The first one is shared by scientific and mythical thinking alike: it is the very existence of semiotic mediations such as linguistic signs and human institutions in general to conceal their role in the construction of reality, be it scientific or mythical. That is the reason why scientific and mythical thinking are being caught in the illusion that semiotic mediations can be disposed of and that reality can be grasped directly "as it is". But scientific and mythical thinking strongly differ in their way of constructing reality through the concealed power of semiotic mediations: whereas scientific thinking builds a limited knowledge in specific domains expanding through history, mythical thinking keeps an all-encompassing viewpoint on reality that keeps it close to the very source of sense-making but does not develop through time. Philosophy appears therefore as a specific knowledge that takes distance from the illusion of the passive role played by semiotic mediations and reflects on their role in the construction of reality, be it mythical or scientific.

Keywords General epistemological project of symbolic practices · Meaning and sense · Positive and productive knowledge

The last chapter has enabled to establish three results of a general significance which correspond to the three historical stages of the analysis of the development of modern physics such as described by Cassirer.

1 Three Learnings from the Preceding Chapter

The study of the revolution having affected modern physics since the Renaissance has shown that the language sciences have had a true influence on the natural sciences albeit mostly a covert one because it is indeed a philological revolution in the usage of signs which has made possible the capital transformations in the interpretation of nature described by Cassirer. It can therefore be said that it is the *usage of signs*

© Springer Nature Switzerland AG 2020
J. Lassègue, *Cassirer's Transformation: From a Transcendental to a Semiotic Philosophy of Forms*, Studies in Applied Philosophy, Epistemology and Rational Ethics 55, https://doi.org/10.1007/978-3-030-42905-8_3

which, at the same time, establishes a relationship between the sciences of language and the natural sciences, while also introducing a fundamental tension within this relationship: in the natural sciences indeed, the usage of signs remains entirely subordinate to the clarification of the signification of *concepts* and must efface itself *entirely* all the while being *reusable at will* in order to claim objectivity whereas that, in the language sciences, the usage of signs is not only subjected to a role of available tool aiming for full conceptual clarification because the effects of signs are, on the contrary, conceived as *always susceptible of continuing to play a role*, even a covert one. It is therefore *the greater or lesser transparency of signs* which introduces a tension between the two poles and which enables to better account for the too rigid distinction between the natural sciences and the sciences of language.

The study of the revolution having touched modern physics has also shown that, in Cassirer's view, what needed to be understood by "progressive diffusion of a form" in a science consisted, in this specific case, in the diffusion of the *functional interpretation*. From an epistemological point of view, the trajectory of modern physics indeed tends to progressively make all instances of judgment functional, even at the root of intuition and of its fundamental categories—space and time—in the theory of relativity.

The scientific description of nature can therefore no longer, for Cassirer, be simply founded on a theory of "correspondence" between concept and nature based on a resemblance, be it only that of the formality of intuition in which the Euclidean character of space plays the role of a phantom of a *mimesis*, as it was in the case of time in which Kantian schematism was conceived on the mode of ideation serving, in a way, as a natural environment for the conceptual description of nature. If space-time is no longer the locus of a relation of resemblance between concepts and nature, then only the conceptual functionalization specific to formalization such as it can be found in a "physics of principles" becomes the locus of this possible relation. But let us be clear. What Cassirer defends, by distancing himself philosophically from Hilbert, is that the conceptual organization made possible by formalization always involves an *anterior non-formalized practice of signs which plays a transcendental role*. Hence the capital importance that Cassirer attributes to symbolism in that it is not reducible to a formalization which keeps with a univocal usage of signs that would be neither inscribed within an interpretative tradition nor likely to receive new interpretations. Formalization must therefore not eradicate the usage of signs as explored by the sciences of language and the epistemological attitude which would consist in strictly keeping with formalized propositions while neglecting all other types of proposition would de facto exclude the field which nevertheless fuels formalization. Any exclusively formal epistemology would thus deprive itself of a part of its own object and would wrongfully reject the reflection on the usage of signs beyond its rational competence. Cassirer's epistemological attitude is quite different: by seeking to reintegrate within the purview of epistemology the practice of language and of signs in general, and this, *in their original conceptual indetermination*, it seeks to found a general epistemology of symbolic practices *which is directly situated at the core of the polarity* existing between the natural sciences and the sciences of language and, more generally, the cultural sciences.

Finally, as found in the preceding chapter, the exemplary place—in a Goethean sense—of the theory of relativity in the development of modern physics comes from that non-Euclidean concepts are for the first time used at the very core of the rational interpretation of nature. The epistemological enigma to which Cassirer is then confronted is the following: how did Einstein manage to shift his point of view and to modify the interpretational framework making possible relativistic physics since he found the Euclidean framework which he inherited to be insufficient when the Euclidean framework only allowed to formulate *problems in Euclidean terms*? How was it possible to change not only the interpreted content, but the *very form* of the interpretation? Beyond the epistemological response to this question regarding the particular case of relativity such as it has already been described above, Cassirer's philosophical response builds on the elaboration of the notion of the symbolic form: *a form is truly symbolic when it possesses itself the capacity to modify the interpretive framework which it nevertheless makes possible.*

To clarify this matter, we should begin with a general remark concerning a particular symbolic form which we will later have the opportunity to examine in greater detail: that of language. Although languages are all equally meaningful and that they do not distinguish themselves from one another in this respect, the study of a single language can by no means suffice in order to experience the nature of language. Language as such can only be apprehended on the basis of the *diversity* of languages and of the acknowledgement of the fact that the diversification of languages forms part of the very essence of language. Moreover, we have a measurement of this diversity, on the one hand from the standpoint of speakers because it is starting with a particular language experienced as being the most familiar (most often one's native language) that the diversity of languages extends from the most familiar to the less familiar and, on the other hand, from a historical point of view because it is possible to produce a general classification of languages and of their evolutions. Now the development of physics may be described in a way in which, without being analogous to the case of languages, may be deemed somewhat similar, owing to the fact that the diversity of geometries finished by being integrated into it.[1] It is indeed possible to keep with the Euclidean framework to describe nature, but this would miss the relation that any interpretative framework should entertain with its other, that is, not only with a particular interpretation—in this case, the Euclidean interpretation—*but with geometrical sense in general*, that is, with the diversity of possible interpretations, Euclidean and non-Euclidean. And there is, in this case also, a measure of this diversity, because it remains on the basis of Euclidean geometry that the usage of non-Euclidean geometries becomes conceivable in physics. The case of relativity is therefore exemplary as it directly opens towards the philosophical problem of *categorical transformations* as they appear in the notion of symbolic form. Two capital epistemological distinctions must be summoned to account for this philosophical issue: the distinction between

---

[1] The analogy is not complete however, since there is no intertranslatability of geometries the ones into the others. For example, it is possible to produce a Euclidean representation of the case of hyperbolic geometry in which there exists an infinity of parallels at a given point using Poincaré's semi-plane, but this remains an exception rather than the rule.

signification and meaning on the one hand and the distinction between positive and productive knowledge on the other. *Neither of these are thematized by Cassirer as such and it is we who introduce them here for the sake of clarity in our exposition.* They are no less capital in order to account for the general project pertaining to a philosophy of symbolic forms.

2   The Distinction Between Signification and Sense

This first distinction is directly involved by the notion of form conceived both as a place of separation but also of exchanges with the environment with respect to which it defines itself. Four remarks appear to be necessary for clarifying the distinction in question.

We should first stress that Cassirer's vocabulary at times suggests that he could have thematized the distinction as such inasmuch as the two terms which are 'signification' and 'sense' often co-occur within a same sentence, underscoring both the necessity of distinguishing them without however making their distinction clearly determinate.[2]

When adopting the standpoint of a particular symbolic form, we realize that it tends to *integrally* occupy a whole order of signification.[3] Now, there are two ways to integrally circumscribe a domain: either by adopting an outside point of view or by positing oneself from within and by studying its viable variations,[4] which gives an idea of the perimeter of the domain in question. This makes it possible to assert, for example, that language is a symbolic form. Not that it would be truly possible to take an exterior stance, but it is rather from within that the "language" form

---

[2]Cf. E. Cassirer, *Axel Hägerström. Eine Studie zur Schwedischen Philosophie der Gegenwart*, ECW 21, p. 87: "But even if it would be the case, only this particular starting point of the conceptions of law would therefore be described and this would not affect the whole of their possible "meanings". Consequently, the constant transformation of this meaning and the motifs that become effective in it constitute the true problem. Hägertsröm repeatedly explains that already in the classical Roman jurisprudence the original mythical meaning had obscured the precise concepts of the law. He differentiates between a "original meaning" of the concept "jus", "justum", "injuria" and its "later meaning". According to him, the original meaning of the concept "jus" is the supernatural force which is characterized by its purity from the stain of mortal germs, and which is also a mythical force of life."

[3]E. Cassirer, *The Philosophy of Symbolic Forms*, vol. 1, p. 81 (ECW 11, pp. 10–11). Another example which Cassirer gives concerning this conflict is the way each symbolic form tends to make itself into an absolute within the cultural world; E. Cassirer, "Form and Technology", in *Ernst Cassirer on Form and Technology; Contemporary Readings*, Hoel A.S. & Folkvord I. (eds.), Palgrave, Macmillan, 2012, p. 42 (ECW 17, p. 173): "Moreover, as technology unfolds, neither does it simply place itself next to other fundamental mental orientations nor does it order itself harmoniously and peacefully with them. Insofar as it differentiates itself from them, it both separates itself from them and positions itself against them. It insists not only on its own norm, but also threatens to posit this norm as an absolute and to force it upon the other spheres."

[4]The case of logic in which it is possible, using a theorem of impossibility resting upon proof by contradiction, to delimit from within a specific conceptual domain also belongs to this type.

reverberates and becomes accessible because it is before anything else *a* language *and* the recognition of the diversity of languages which enables to conceive of the nature of language in general as a *particular* order of signification. Likewise, it is possible to conceive of technology as a symbolic form if we conceive of the *variation* of a particular technology as the modality by which any technology manifests the specific order of signification pertaining to it.

Regarding the relationships between symbolic forms, it must be recognized that, all the while occupying a specific order of signification, symbolic forms nevertheless present reciprocal relationships of distance and proximity *from the standpoint of sense*. There is therefore a *signification that is internal* to a particular symbolic form, a signification which is requited through the usage of a particular symbolism and a *sense*, more diffuse, which stems from what is exterior to the form in question. If we compare two particular forms such as language and technology, we indeed see that technology conceived as a symbolic form is neither opposed to nor identified with language and that it is both possible to technically equip a language, for example by using a written lexicon and grammar, all the while considering oratory art as a specific linguistic technical tool. There is therefore no hermetic border between forms because the modalities of the expression of sense can intersect or, on the contrary, distinguish themselves from one another. Also, all symbolic forms evolve historically in their own area of signification because they have access to the other forms of which they apprehend from outside the significations as senses. It is because the associated relation of signification and symbolism—the specific usage of signs— is never fully stabilized to an extent where the symbolism is strict enough to not be interpreted differently than it is: hence the possibility to *grasp sense* by means of a specific usage of symbolism and not only to *set significations*. We will therefore use this distinction between signification and sense in order to clarify the distinction between a form considered from the inside and a form in its relation to other forms, even if the distinction between signification and sense has not been thematized by Cassirer using such terminology.

Finally, we must note the particular place occupied by philosophy which *is not a symbolic form* because it is not circumscribed within one of the specific orders of signification but because it is rather the possible locus of their reciprocal reverberation. While it is close to the inherent capacity of language to speak of all forms, it distinguishes itself by its adoption of a *conceptual* type of knowledge and by the critical distance it takes with respect to the usage of signs which is neither reducible to that of science, neither to that of myth—the two extremities identified by Cassirer in the usage in question. From this standpoint, a *philosophy* of symbolic forms can find inspiration in what is elaborated in particular symbolic forms without limiting itself to it. Thus, the exemplarity of relativistic physics as a conservative extension of Euclidean physics indeed contributes in conceiving the idea of a philosophy of symbolic forms inasmuch as the specific case of relativity *makes sense* for philosophy, but this same particular case transmutes within it into philosophical *signification* which is directed towards a completely different horizon where the "lacking" conceived as a quest for universality, as we have called it above,[5] is made into an actual theme.

3   Distinctions Concerning Knowledge

The *princeps* epistemological distinction which Cassirer uses to distinguish ancient and medieval science from modern and contemporary science, that is, the distinction between *substance* and *function*, intersects with another one which is less apparent in Cassirer's work but which is no less crucial with respect to the general architecture of a philosophy of symbolical forms: the distinction between *positive knowledge* and *productive knowledge*. Cassirer indeed remarks that even scientific knowledge begins to be conceived as an *immediate* relation to "reality" conceived independently from the view cast upon it and that from this perspective, it is not on such a basis that it distinguishes itself from the other form of knowledge with which it is in highest contrast: mythical knowledge.[6] Scientific knowledge indeed also rests at first on the *illusion* consisting in believing that it immediately describes "reality" and it is only progressively that it departs from this illusion by recognizing the role of symbolic mediation specific to signs.[7] Scientific knowledge thus defines the possibility for a history which is *the history of its divorce from immediateness* and the *recognition of the primacy of the mediation by signs*. From this point of view, if scientific knowledge is opposed to mythical knowledge, it is not in terms of content as such which, in both cases, always aims to provide an explanation of phenomena, but because scientific knowledge is susceptible of *partially departing form this condition of immediateness*, a departure of which mythical knowledge is precisely incapable and which characterizes it, at least negatively. This is the reason for which Cassirer speaks of the substantial form of scientific knowledge, as it was conceived of during the ancient and medieval periods, as a "semi-mythical" form: substantially conceived scientific knowledge was indeed mythical in its illusion of an immediate relation to "reality" but it was *not solely* mythical—only "semi-mythical"—inasmuch as the possibility of its divorce from the immediateness also defined the conditions for the inception of its own renewal, contrary to mythical knowledge. What characterizes the exemplarity of scientific knowledge is therefore the fact that it is, in Cassirer's view, in the utmost manner *capable of spurring the evolution of the interpretive framework*

[5]Cf. Chapter 1, Conclusion.

[6]E. Cassirer, *The Philosophy of Symbolic Forms*, vol. 2, p. 26 (ECW 12, p. 33): "In knowledge, too, the *use* of hypotheses and principles precedes the *knowledge* of their specific function as principles—and until this insight is gained, science can only contemplate and state its own principles in a material, that is, semimythical form."

[7]E. Cassirer, *The Philosophy of Symbolic Forms*, vol. 2, p. (ECW 12, p. 33): "For what distinguishes science from the other forms of cultural life is not that it requires no mediation of signs and symbols and confronts the unveiled truth of "things in themselves" but that, differently and more profoundly than is possible for the other forms, it knows that the symbols it employs are *symbols* and comprehends them as such."

which it founded itself, as we have already stated numerous times. Scientific knowledge is therefore intimately linked with its historical progression, contrary to a naïve epistemological attitude considering on the contrary that the specificity of scientific knowledge consists in liberating itself from history. What is then exemplary with scientific knowledge is that it opens onto a truly philosophical dimension, that of *productive knowledge* in general which consists in the capacity of operating an evaluation of the mediation role specific to signs *in all types of knowledge*. There is therefore a positive knowledge pertaining to *science* as there can also very well be a positive knowledge of *myth* which is then reintegrated into the field of rationality, in its own particular place, that of the lowest degree of recognition of the mediatory role of signs. Scientific knowledge then distinguishes itself from mythical knowledge by the attitude which is its own with respect to the awareness of the role played by the mediation by signs: contrarily to mythical knowledge, resolutely ahistorical according to Cassirer, scientific knowledge is historical in the sense that it is founded upon an accumulation of elements which always remain available and which can always be reworked, independently of the age during which they came to light. We have also seen in the previous chapters the two aspects specific to scientific knowledge of which it is question here. The first concerned the existence of an immediate and therefore "semi-mythical" relation, to use Cassirer's expression, between Aristotelian logic and the Greek language.[8] The second concerned the profound renewal of geometry over the course of the nineteenth and twentieth centuries which is founded on the employment of Euclidean elements remaining available for re-elaboration by the mathematical tradition.[9] Knowledge therefore appears as a positive, never fully established process, *including in scientific knowledge* and must not be confounded with what we call "productive knowledge" nor be considered as the final term of the process of mediation which signs make possible. *Productive knowledge, for its part, pertains to philosophy*, and it is owing to philosophy alone that it is possible to distinguish different modes of knowledge in function of the type of attitude which they develop with respect to signs.

---

[8]Cf. Chap. 2, § 11.
[9]Cf. Chap. 1, § 13.

Chapter 4
Introduction to the Notion of Symbolic Form

Abstract This chapter deals with Cassirer's central notion of a "Symbolic form". Acknowledging the fact that nowhere does Cassirer give a clear definition of what a "Symbolic form" is, the chapter starts by dismissing two misconceptions that would too reductively label a "Symbolic form" either a conceptual framework (like Darwinism or formalism) or a methodological tool (like differential calculus or perspective). Rather, "Symbolic forms" are viewed as instituting powers that enable different modes of meaning production by inducing distinct semiotic perspectives on a symbolic material. This dynamical view has two far-reaching consequences on the way of understanding the notion of a "Symbolic form". Firstly, although Cassirer himself had difficulties departing from the idea that Language and Myth were stepping stones towards the later accomplishment of conceptual knowledge, their role should be understood otherwise so that his later work on other "Symbolic forms" like Art, Law or Technology would be clarified. As a consequence, any "Symbolic form" should rather be viewed as an example of a stabilized semiotic perspective on a symbolic material. Cassirer, although shaky in his vocabulary, describes these various semiotic perspectives by using three operators he describes as Expression, Evocation and Objectification that are to be found in all "Symbolic forms" whatever their number. Stabilized "Symbolic forms" emerge at the intersection points of these operators.

Keywords General characterization of a symbolic form · Ambivalence of language in its foundational role · Symbolic forms as effects of trans-categorical operations

© Springer Nature Switzerland AG 2020
J. Lassègue, *Cassirer's Transformation: From a Transcendental
to a Semiotic Philosophy of Forms*, Studies in Applied Philosophy, Epistemology
and Rational Ethics 55, https://doi.org/10.1007/978-3-030-42905-8_4

Cassirer does not present the notion of symbolic form using a definition, even if the notion is of course present in all of his works.[1] Therefore, it is necessary to directly delve into his work in order to better circumscribe where lies the core of his philosophical project, a glimpse of which was given in the two previous chapters aiming more specifically to describe the particular case of the natural sciences during the modern era. This is because all of Cassirer's analysis pertaining to the particular case of modern physics led to consider science as a symbolic form among others, even if the quasi-Promethean project of such science appeared, during the seventeenth and eighteenth centuries, able to occupy the whole field of rationality.

1 What a Symbolic Form Is Not

First, two misunderstandings concerning the epistemological aspect of the notion of "symbolic form" should be dispelled.

On the one hand, indeed, a "symbolic form" is not a "theoretical framework", a "discourse about" such or such type of activity, such or such type of fact, or such type of knowledge. More specifically, a symbolic form is *not only* an epistemological framework which would enable to characterize, using a method defined *a priori*, a certain type of theoretical object, as are, for instance, "Darwinism" with respect to the object which is the 'evolution of species', or "formalism" with respect to the object which is 'finitist axiomatic systems'. Any epistemological framework participates however to the notion of symbolic form but what appears more profoundly in it is the *instituting power* of the symbolic form, one which *founds* a certain manner of producing signification and of apprehending in this way objects and interactions which would not otherwise have had a determined existence. This is the case with the first examples of symbolic forms described by Cassirer in his book *The Philosophy of Symbolic Forms*: language, myth, and science, which do not, epistemologically speaking, all have the same status, but which all participate nevertheless to the notion of symbolic form inasmuch as they *institute* a *certain type* of presentation of sense in the form of a certain mode of signification. The notion of symbolic form thus enables before anything else to specify what must be understood by the *effectivity* specific to signs when they *institute* a certain specific way of making sense. It is towards this instituting power that the notion is first directed and of which some idea is given

---

[1]Only the article "The Concept of Symbolic Form in the Construction of the Human Sciences" addresses the question more directly, even if no definition, strictly speaking, is proposed in it. What comes the closest is the following passage, *Ernst Cassirer, The Warburg Years (1919–1933)*, p. 76 (ECW 16, p. 79): "By "symbolic form", one should understand every energy of spirit by which the content of spiritual signification is linked to a concrete and intrinsically appropriate sensuous sign. In this sense, language, the mythical-religious world, and art confront us as particular symbolic forms. For in each of them the basic phenomenon takes shape; our consciousness does not content itself with receiving impression from the outside, rather it links and penetrates every impression with a free activity of expression. Thus a world of self-created signs and images emerges that opposes and asserts itself in independent fullness and original force against which we designate as the objective reality of things."

when an epistemological framework enables to produce a specific object field but to which the notion of symbolic form is not limited because it is the *production instituting such frameworks* which constitutes its specific nature.

On the other hand, a "symbolic form" is not only a tool or device that one may elect to use or not. Its instituting power envelops a *necessity* which makes it a mandatory transition point for all activities of a certain type[2]: thus, the linguistic nature of human communication is what it is *and not otherwise*. Likewise, mythical construction is what it is *and not otherwise*, contrary to all attempts which would seek to "straighten up" language to make it more perfect, and contrary to any critique of mythology which would see in it only superstition. The matter of concern here is therefore not only the fact any activity pertaining to a symbolic form is made possible by internal rules, grammatical ones in the case of language, but also that it encompasses a necessity of another order *imposing* its usage. We could see here the effect of a vicious circle: if we define an activity (for example, speaking) by the necessity instituted by the form which makes the activity possible (in this case, language) and the necessity of the form in question by the activity of speech which it enables to deploy, we will indeed not have progressed in the characterization of the specific necessity of the symbolic form which is that of language, other than attributing a name to an activity. But it is not a vicious cycle when the true nature of the notion of form is understood. From an internal perspective, a form envelops its own clarification in the transformations which affect the necessity of its usage: thus, for example, the pathologies which affect language as they are studied by Cassirer in the third volume of *The Philosophy of Symbolic Forms* enable to measure how it is in the case of normal linguistic communication, deployed according to its own necessity; likewise, the mandatory usages within styles and contents—be they linguistic, mythical, artistic, or scientific—as well as their possible variations constantly sought by subjects give an idea of the ever to be constructed necessity which inhabits them. This necessity then takes the aspect of the quest for that which could provide *authority* regarding the production made possible by a given form and there is no vicious circle in stating that a form is what it is and *not otherwise*, if we understand this "not otherwise" as always remaining to be justified. From an external perspective, the multiplicity of symbolic forms, their kinship, their distance, and endurance, shows that it is possible to adopt regarding each of these relations an external point of view enabling to measure that which, in the production specific to each, manifests its idiosyncratic necessity. Thus, it is possible for example to see what language *is not* when taking the standpoint of another form such as technology or science and to better circumscribe that which makes its specific norm.

---

[2]It is in this sense that the instituting power of a symbolic form inherits that which stemmed from the necessity specific to what the critical tradition linked to the transcendental, that is, the necessary conditions of possibility of the objects of experience.

2 What It Is

It must first of all be noted that the instituting power which is specific to symbolic forms may be disconcerting in particular for those expecting an analysis remaining in strict keeping with the tradition of Kantian criticism, because the *a priori* conditions of possibility of the objects of experience no longer manifest themselves via abstract categories, but rather by certain *signifying forms* which are *perceptible in objects* (a language, a ritual, a type of equation), all the while governing the way in which to produce sense and signification—in short, forms which play a role that classical criticism would have deemed transcendental. As we have already noted, this marks for the critical perspective a turning point which could be qualified as "Goethean": the difficulty and the criticisms it is likely to provoke come from the almost automatic habit consisting in supposing a strict division between that which pertains to the categories of a "subject" and that which would pertain to an "objective nature" which would preexist, whereas the notion of symbolic form would reside prior to this division and would seek to show how this distinction constructs itself in the diversity of forms. On the contrary, the notion of symbolic form has the originality of allowing view the effectivity of signs as per the production of sense by individuation, competition, and distancing between forms and not by means of a fixed categorical framework. This is why a symbolic form is never isolated, but must always be understood with respect to the process which likens it or distinguishes it from one or several others. Thus, for example, the movement of distancing accomplished by the symbolic form which is science in its relation to the more primitive symbolic form which is myth may not be grasped without science being put at the same time in relation with its mythical origin: from this point of view, science is not in opposition with myth as if they had nothing in common, but it is indeed, on the contrary, *because they have something in common*—in particular the notion of causality, as we will later see—that they compete and conflict with one another. This is what enables to understand in particular that science participates in the instituting effectivity of symbolic forms in a *particular* place when it is only put into relation with other forms but it nevertheless occupies an *eminent* place when it is the process of distancing with respect to other forms which is emphasized. Sense thus appears under various modes of effectivity and is called "language", "myth", or "science" in function of the relation it has with its own internal diversity which manifests as a diversity of forms. What Cassirer attempts to describe philosophically is therefore the dynamic process pertaining to the relations of conflict or coincidence between forms: being shaped *for itself*, a form becomes autonomous with respect to another but then develops the specific form of its signification *as if its norm was the sole one* and as if[3] it could by itself *saturate the field of signification*, thus generating conflicts and covert similarities with other forms which participate in the transformation and emergence of new relations between forms. In fact, there exists an inherent tendency in all symbolic forms aiming to occupy a place which *does not exclusively belong to*

---

[3]Cf. supra, § 12.

them,[4] which necessarily instills *conflict* between forms. Where does this tendency come from? Cassirer says little about this, but it is possible to suppose that it comes from a *confusion between signification and sense* having for origin the role conferred to what is meant by 'symbol'. If the instituting productivity specific to a symbolic form is not likely to make its relation to its own mode of signification evolve, a confusion will introduce itself between sense and signification that will undermine the necessity for a *constant re-elaboration of the norm* and therefore also the very idea of the diversity of forms, be it internal or external. From this results antinomies from which the notion of symbolic form must enable to steer away by favoring the diversity of meaning of those forms. Otherwise, these forms would tend to wither due to a too strictly limited usage of their relation to the symbol in an illusory attempt to appropriate the totality of the field of meaning on the mode of signification.

There lies a profound inner transformation of the Kantian perspective inasmuch as it was indeed the signification specific to science, once conceived from a critical viewpoint, which enabled it to escape the register of the transcendental illusion whereas here, science, like any other form, would be on the contrary plunged into illusion if it were to keep only with the simple determination of signification in its own order. Conceiving the nature of science as precisely as possible is therefore capital for Cassirer's undertaking because it is finally this reflection which governs the manner in which to conceive the very process of construction of the relation between sense and signification. From this point of view, keeping with the possibility of a total elimination of other symbolic forms by science would amount to continuing to use them unconsciously, that is, in the most offhanded manner possible, and in transforming science into a dogmatic undertaking of a purely mythical nature. On the contrary, if we recognize symbolic forms as having a status of *particular presentations of sense* each deploying internal significations, be it within the mode of *object constitution* in the natural sciences or of *sign institution* in the cultural sciences, science will then be restituted within a new field which it does not entirely cover by itself, without however its own specificity being underestimated in any way. We then understand that Cassirer came to describe the notion of symbolic form in the following manner:

> Each of the original directions of knowledge, each interpretation, which it makes of phe-
> nomena to combine them into the unity of a theoretical connection or into a definite unity of
> meaning, involves a special understanding and formulation of the concept of reality. There
> result here not only the characteristic differences of meaning in the objects of science, the
> distinction of the "mathematical" object from the "physical" object, the "physical" from
> the "chemical", the "chemical" from the "biological", but there occur also, over against the
> whole of theoretical scientific knowledge, other forms and meanings of independent type
> and laws, such as the ethical, the aesthetic "form". It appears as the task of a truly universal
> criticism of knowledge not to level this manifold, this wealth and variety of forms of knowl-
> edge and understanding of the world and compress them into a purely abstract unity, but to
> leave them standing as such. Only when we resist the temptation to compress the totality of

---

[4]We could think, for example, of a language which indeed occupies all the field of linguistic signification in its form and which does not let itself be used as a basis upon which to conceive of the diversity of languages, an element which is nevertheless indispensable for thinking about language in general.

forms, which here result, into an ultimate metaphysical unity, into the unity and simplicity of an absolute "world ground" and to deduce it from the latter, do we grasp its true concrete import and fullness. No individual form can indeed claim to grasp absolute "reality" as such and to give it complete and adequate expression. [...]. It is the task of systematic philosophy, which extends far beyond the theory of knowledge, to free the idea of the world from this one-sidedness.

A capital change is initiated in this 1921 text.[5] It is necessary to refrain from conceiving the relation to the object in general exclusively according to the mode of the adequation of an intuition to an object, an issue which remains captive of the notion of resemblance because it is the issue of the adequation to the object which leads to the quest for the univocal constitution of signification. Now, this fascination for univocity, implying that the considered object is conceived as being already fully determined *before* any construction, is indeed made possible by a complete reversal of the *conditions of access* to the object: what is *terminal* (the constructed objectivity of signification) is conceived as always having been *already there*, but on the mode of the still abstract essence to be uncovered. This is the point of view from which Cassirer managed to radically distance himself[6]: what is "already there" does not yet have a specific sense which would be uncovered based on its logical essence, but there is, at any moment in time, a *sense-generating plurivocity* of which the objectivity manifested in science is but a modality. It is therefore indeed a matter of broadening the *a priori* conditions of the presentation of meaning to all activities which convey a specific mode of signification. The notion of "symbolic form" will be put into the service of this end.

It can therefore be stated that for Cassirer, an activity is said to be "symbolic" when its progression is shaped by a norm that involves a particular modality in producing signification specifically characterizing this activity all the while preserving the possibility of transforming this modality by constantly re-elaborating its relation to the norm, that is, through a play on sense. This specific usage of instituting production then deploys a flexible form which is likely to extend in function of the norm pursued and to be modified over the course of its own activity. Thus, the recourse to a norm, in the shape of a requirement which transcends the present situation anticipates

---

[5]E. Cassirer, *Einstein's Theory of Relativity*, pp. 446–447 (ECW 10, pp. 112–113).

[6]E. Cassirer, "The Object of the Science of Culture", in Ernst Cassirer, *The Logic of the Cultural Sciences*, Yale University Press, pp. 29–30 (ECW 24, p. 386): "For once we are convinced that the logical concept is the necessary and sufficient condition for the knowledge of the essence of things, then everything else, which is specifically different and which does not meet this standard clarity and distinctness, ends up being only an unreal appearance. In this case the illusory character of those spiritual forms that remain outside the sphere of the purely logical cannot be contested; it can, as such, be demonstrated, and in this respect explained and justified, only to the extent that we inquire into the psychological origin of the illusion and attempt to illustrate its empirical conditions in the structure of human imagination and human fantasy. However, the question takes a completely other turn if, instead of treating the essence of things as having been fixed from the beginning, we see in it, on the contrary, the infinitely distant point toward which all knowing and understanding aim. In this case, the « given » of the object is transformed into the « task » of objectivity. And this task, as it can be shown, does not involve theoretical knowledge alone; rather, every energy of spirit participates in its own way."

the future evolution of the form: the form is then said to be "symbolic" because in conforming to the requirements of a norm, it institutes a field in which the relation between signification and sense manifests a particular modality of the productivity of forms. Thenceforth, the symbols of a given field appear as the incarnation of this relation, unstable by nature, between signification and sense. By relating the notion of symbolic form to *energeia* (activity) in contrast to *ergon* (task),[7] Cassirer considers that it becomes possible to conceive of the institution of signs and of the constitution of nature, not in order to establish a false continuity between the two but on the contrary in order to study the possibility of their kinship or rupture.

It then becomes possible, from this first characterization, to specify the broad directions taken by the idea of the presentation of specific relations between signification and sense. Originally, as we have already seen, *The Philosophy of Symbolic Forms* of which the three volumes were published between 1923 and 1929 distinguishes three relations of this type: language, myth, and science. But others are later added in Cassirer's work: ethics, aesthetics and law, examples of which have often been addressed by Cassirer without them having always been the object of a systematic exposition. There is therefore an evolution in Cassirer's thought which poses a certain number of difficulties with respect to the global coherence of his project, as we shall now see.

3 The "Symbolic Turn"[8]

In their original form—as they present themselves in *The Philosophy of Symbolic Forms*—symbolic forms are *modes of knowledge* (of which the opposing poles are myth on the one hand and science on the other) which distinguish themselves by their greater or lesser proximity with respect to the fundamental genre of *signifying activity*, language. Hence the trilogy of forms having already been mentioned on multiple occasions: the *activity* of language (volume 1) serving as a foundation for the most opposed types of *knowledge* which are myth (volume 2) and science (volume 3). The relations between these three forms must enable to account both for the possibility of philosophical knowledge as the acquisition of an awareness of the mediation of signs and as justification of this productive thinking. As seen earlier, it is useful to distinguish two points of view concerning the mediation of signs: that of signification and that of sense, of which the determination can be enriched here by characterizing signification as "immanence" and sense as "horizon". What

---

[7]Cassirer returns here to the distinction highlighted by von Humboldt who, for his part, had borrowed it from Aristotle. Cf. for example, E. Cassirer, "Structuralism in Modern Linguistics", *Word*, 1, August 1945, p. 110 (ECW 24, p. 310). In the same line of thinking, Cassirer also uses the distinction between *forma formans* and *forma formata* in 'Form and Technology' p. 18 (ECW 17, p. 142) and in 'Geist und Leben' (ECN 1, p. 18) or *ordo ordinans* and *ordo ordinatus* in 'Grundproblem der philosophische Anthropologie' (ECN 1, p. 99).

[8]The expression "symbolic turn" is attributable to John M. Krois in "The priority of "symbolism" over language in Cassirer's philosophy", *Synthese*, 2008.

should be understood by "immanence" is the manner in which each symbolic form deploys itself in its own autonomy and poses for itself its internal principles of legality.[9] By "horizon", what should be understood is the way in which forms have incidences upon one another, be they in the form of convergences or of ruptures. Only philosophical knowledge is capable of articulating immanence and horizons by studying the incidences in question because they suppose a point of view which both transcends all particular forms all the while basing itself on an immanence which is only of a "critical" nature, that is, without becoming itself the objective and ultimate end to which all symbolic forms would only partially tend.

A difficulty nevertheless presents itself due to the fact that the original ternary schema of symbolic forms has evolved over the course of time in Cassirer's work and that it found itself muddled by the fact that other symbolic forms had come into play, either as types of activities (for example, technology), either as modes of knowledge (for example, art). How can these new forms be integrated to the already constituted set without putting into question the equilibrium between a signifying activity and modes of knowledge? It may indeed be considered that it is precisely the role attributed to the notion of *diversity* of forms which enables to transition from the one to the other, but this would entail a double risk: on the one hand, *loosing sight of the complementarity* of views of signification as immanence and sense as horizon, and on the other hand *accumulating without order* any sort of signifying activities (for example, money used as a currency which pertains neither to language nor to technology) and of modes of knowledge (for example, literature which belongs neither to myth nor to science). What interest then would there be in using the notion of "symbolic form" to designate such disparate realities?

---

[9]E. Cassirer, *The Philosophy of Symbolic Forms*, vol. 1, p. 177 (ECW 11, p. 122): "In defining the distinctive character of any spiritual form, it si essential to measure it by its own standards. The criteria by which we judge it and appraise its achievement, must not be drawn from outside, but must be taken from its own fundamental law of formation. No rigid "metaphysical" category, no definition and classification of being derived elsewhere, however certain and firmly grounded these may seem, can relieve us of the need for a purely immanent beginning. We are justified in invoking a metaphysical category only if, instead of accepting is as a fixed datum to which we accord priority over the characteristic principle of form, we can *derive* it from this principle and understand it in this light. In this sense every new form represents a new "building" of the world, in accordance with specific criteria, valid for it alone."

This difficulty related to the very architecture of *The Philosophy of Symbolic Forms* as been highlighted by several authors[10] and we have already encountered it on several occasions. Two philosophical interpretations of this may be contemplated.

The first consists in favoring as *final end* the self-revelation of the transcendental subject, which would have two mirrored consequences: language would become the immanent *and* transcendent starting point of this self-revelation and consciousness would be its terminal locus.[11] This would be a classical way of interpreting the relation between the immanent point of view and the transcendent one, a relation of which the gnoseological foundation would be the *conceptual* determination taking root in language.

The second would concern the exploration of the *semiotic specificity* of each symbolic form without favoring language in particular and while trying to interpret the self-revelation of the transcendental subject not as the gradual progression of consciousness towards itself but as the exploration of semiotic modes of expression such as they manifest in symbolic forms conceived above all as a public material. These forms are *collective from the onset and partial but undefined*, and they directly have a social value owing to the constant work of re-elaboration which they operate upon their own norm. This public material would need to be constantly reworked by subjects who, through such activity, would bring symbolic forms into existence, all the while, by the intermediary of these very forms, would experience their own sociality in an essential manner. In this would lie another way of articulating immanence and horizon from the standpoint of philosophical knowledge inasmuch as, supposing that symbolic forms are social in nature, there would no longer be a need to consider access to productive knowledge as a univocal itinerary necessarily leading to the subject's full self-knowledge. Let us address this alternative in more detail.

---

[10] As L. Iribarren stresses in "Langage, mythe et philologie dans la *Philosophie des formes symboliques* d'Ernst Cassirer" (*Revue germanique internationale*, 15/02, pp. 95–114), this point has been emphasized on several occasions, particularly by Blumenberg and Habermas. For Iribarren, the disparity inherent to the aims which Cassirer set out in *The Philosophy of Symbolic Forms* stem from the fact that Cassirer seeks both to establish the autonomy of symbolic forms all the while considering the succession of forms in a finalized trajectory of consciousness towards scientific knowledge.

[11] E. Cassirer, *The Philosophy of Symbolic Forms*, vol. 3, p. 78 (ECW 13, p. 87): "The world of the spirit forms a very concrete unity, so much so that the most extreme oppostions in which it moves appear as somehow mediated oppositions. In this world there is no sudden breach or leap, no hiatus by which it breaks into disparate parts. Rather, every form through which consciousness passes seems to belong in some way to its enduring heritage. The surpassing of a particular form is made possible not by the vanishing, the total destruction, of this form but its preservation within the continuity of consciousness as a whole [...]."

3.1 The Ambivalence of the Place of Language
in the Architecture of Symbolic Forms

The first attempt at a solution would consist in assigning to language a central place among symbolic forms[12] and to then see in it a matrix based on which it is possible to conceive both the appearance of new forms and their coupling with language—the paradigmatic example of such coupling being Cassirer's study of the relations between myth and language in *Language and Myth* (1925).[13] However, this solution, oscillating between mentalism and objectivism, was in fact partially defended by Cassirer's teacher, Paul Natorp.[14] Cassirer rejects neither its interest nor its relevance with respect to particular analyzes and he actually borrows its general lines when seeking to clarify the relation between language and consciousness on the one hand,[15]

---

[12]Cassirer historically attributes this solution to Herder and Humboldt; E. Cassirer, "Form and Technology" p. 36 (ECW 17, p. 166): "In German intellectual history, this purely aesthetically composed and grounded 'humanism' gradually grew, and how another cultural power locates itself, independently and equally, next to art. For Herder and Humboldt it is language that shares with art the role of creator and seems to be the basic motive for the real 'anthropogeny'."

[13]This is the solution advocated by L. Iribarren in the cited article, which follows Habermas regarding this point: "Then of course Cassirer would have been placed in a position where he did not want to be: he would have been obliged to transform the only heuristic priority through which the transcendental analysis of language and the linguistically constituted lifeworld was made productive in his research into a systematic priority and to grant a central role to language and to the lifeworld in the construction of symbolic forms". Jürgen Habermas, „Die befriende Kraft der symbolischen Formgebung", in D. Frede and R. Schmücket (eds), *Ernst Cassirers Werk und Wirkung*, Darmstadt, Wiss. Buchges., 1997, pp. 99–100.

[14]E. Cassirer, *The Philosophy of Symbolic Forms*, vol. 3, pp. 55–56 (ECW 13, p. 61): "An obvious difficulty arises as soon as we attempt to fit the concrete whole of the Symbolic Forms into the general framework offered by Natorp's psychology. It is evident that the over-all plan of this psychology must allot an important role to the analysis of language, for determination by the word is an indispensable preparation for determination by the pure concept. And Natorp's psychology, in its program at least, expressly recognized this importance of language. It stresses that an objectivizing power and achievement is contained not only in the scientifically fixated concept or in the scientifically grounded judgement, but in every linguistic sentence. [...]. But in its actual development Natorp's psychology does not do justice to this declaration of principles, for wherever the process of objectification is described, the description is oriented toward the ultimate and supreme phase that is manifested in scientific thinking and knowledge. The very definition of the subject-object relation is derived from this and this alone. The direction of objectivity coincides for Natorp with that of the necessary and universally valid—and this last in turn with that of law. Thus for him the concept of law includes all objectification—regardless of its form and stage". And Cassirer points out a little bit further, p. 56 (ECW 13, p. 62): "And even in the field of theoretical objectification, the role here assigned to the concept of law becomes problematic as soon as the inquiry—as Natorp's own basic principles demand—is directed not toward the concepts of scientific knowledge but toward the concepts of language. For these reveal throughout a form of determination which is by no means identical with determination in and through law. The universality of linguistic concepts does not stand on the same plane as the universality of scientific, and particularly of natural-scientific laws: the one is not merely an extension of the other; on the contrary, they move in different trends of spiritual formation."

[15]E. Cassirer, *The Philosophy of Symbolic Forms*, vol. 1, pp. 268–269 (ECW 11, p. 237): "The problem takes on a new meaning if we reflect that 'things' and 'states', 'attributes' and 'activities'

and the relation between language and logic on the other hand.[16] But it presents at least three shortcomings which seen prohibitive when one considers Cassirer's project as a whole, that is, when considering a philosophy of symbolic forms.

Firstly, it tends to reduce language to an essentially psychological and non-social dimension since it suggests that language enables the self-reflection of *individual* consciousness. Language then risks becoming the *mirror* of *mental* categories, whereas Cassirer seeks to make neither of the theory of reflection nor of an essentially mental dimension the foundation of his undertaking.[17]

Then, it tends to make of the univocity of *logic* the ultimate end conferred to language since it makes possible objectivity in an essential manner and therefore tends to cast aside the symbolic forms described by Cassirer which are not directly coupled with language—at least not in the same respect as myth or scientific knowledge—the first and foremost among these being the activities that are ritual and technology, which Cassirer examined starting in 1925.

Finally, it tends to operate a synthesis of the two preceding points by accrediting the possibility for establishing a *parallel* between a cumulative development of language construction along three stages—analogical, representative, and symbolic[18]—and a cumulative development of symbolic forms throughout history understood as a

---

are not given contents of consciousness, but modalities and directions of its formation. [...]. It is the fixation *into* an object or activity, not the mere naming of the object or activity, that is expressed in the spiritual operation of language as in the logical operation of cognition".

[16] E. Cassirer, *The Philosophy of Symbolic Forms*, vol. 1, p. 268 (ECW 11, p. 147): "It is in the "intuitive forms" that the type and direction of the spiritual synthesis effected in language are primarily revealed, and it is only through the medium of these forms, through the intuitions of space, time and number that language can perform its essential logical operation: the forming of impressions into representations."

[17] Cassirer shows for example that expressivity exists in the animal world without the category of the "mental" as it is usually projected on the human species. Cf. E. Cassirer, *The Philosophy of Symbolic Forms*, vol. 3, p. 93 (ECW 13, p. 85): "But if we examine these observations, they show only one thing with certainty: in how high degree characters of pure expression predominate in the world of animal perception, and how decidedly they outweigh any objective perception of things and attributes. They do not, however, prove that for the animal these characters necessarily adhere to a definite subject, much less to a clearly apprehended person, and that they can only be experienced by way of this vehicle."

[18] E. Cassirer, *The Philosophy of Symbolic Forms*, vol. 1, p. 190 (ECW 11, p. 137): "In general, language can be shown to have passed through three stages in maturing to its specific form, in achieving its inner freedom. In calling these the mimetic, the analogical and the truly symbolic stage, we are for the present merely setting up an abstract schema—but this schema will take on concrete content when we see that it represents a functional law of linguistic growth, which has its specific and characteristic counterpart in other fields such as art and cognition."

self-revelation of the subject as per a mode inspired by Hegel.[19] Now, language cannot serve to conceive any *cumulative* self-revelation of the Subject even throughout history by means of the sole vector of logic because *language resists any unilateral confinement* into a pre-established order[20] which would necessarily conduce it towards the univocity of scientific knowledge and the subject's self-revelation: it is because language *indefinitely* maintains a kinship with myth that it cannot serve as vector for this self-revelation and that it cannot integrally abolish itself in scientific knowledge governed by the univocity of signification, even if such univocity can indeed be produced locally. In this situation specific to language, a major anthropological invariant must be recognized: the meaning of any word or sentence seems indeed to exist in language before its usage by a speaker in a specific occurrence and what is described as "language" appears in this perspective as both a *reservoir* of meaning, a *power* to signify, and an *antecedence* with respect to any speech which would confer it an *originative power that is mythical from the onset* and of which no speaker can be considered to be free during the act of speech. Now, obviously, nowhere does this "language" exist and the speed with which a language transforms to the point of becoming opaque to speakers separated by only a few generations suffices to measure the *profoundly mythical* aspect of such a relation to language, an aspect which is *nevertheless indispensable* to any speech act.

In other words: either symbolic forms constitute a differentiated unit and never disappear in the always possible incidence they may have upon one another. But then, the univocal self-reflection of consciousness would not be the ultimate goal attributed, through language, to philosophical knowledge. Or on the other hand, the symbolic form of myth can indeed be separated from that of language and the univocal self-reflection of consciousness can become the ultimate task assigned to language and, by extension, to any symbolic form. The difficulty will then be the following: on the one hand, it seems impossible in Cassirer's view for a symbolic form (be it myth or another form[21]) to fully disappear to the benefit of the others and for a privileged direction of the type of logical univocity to then be assigned to language; on the other

---

[19] E. Cassirer, *The Philosophy of Symbolic Forms*, vol. 3, p. 78 (ECW 13, p. 88): "The world of the spirit forms a very concrete unity, so much so that the most extreme oppositions in which it moves appear as somehow mediated oppositions. In this world there is no sudden breach or leap, no hiatus by which it breaks into disparate parts. Rather, every form through which consciousness passes seems to belong in some way to its enduring heritage. [...]. "The life of the actual spirit", writes Hegel in this connection, "is a cycle of stages which on the one hand still subsist side by side and only on the other hand appear as past. The features which the spirit seems to have left behind it are also present in its depths" (Hegel, "Vorlesungen über die Philosophie der Geschichte", *Sämmtliche Werke* (Leipzig, 1949), 9, 98)". The consequence of such a point of view has been noted almost in passing by B. Recki in "Cassirer and the problem of language" in "Cultural Studies and the Symbolic" ed. By Paul Bishop and R. H. Stephenson, Northern Universities Press (2003) when she remarked p. 5: "Of course, this schema of mental development poses a problem, insofar as the three developmental steps of disposition over abstraction which are posited in what we may see as Cassirer's approach to a philosophy of history are also seen systematically as three elements in the functional structure of language."

[20] Cf. infra 'Idiomatic Expression in Language' § 'Perception under the Aegis of Language'.

[21] E. Cassirer, *The Philosophy of Symbolic Forms*, vol. 3, p. 78 (ECW 13, p. 88): "If this view is sound, we shall not be able to believe that even so strange and paradoxical a structure as mythical

hand, it is also clear, for Cassirer, that the appearance in language of a symbolic mode such as the univocity of the logical and, more generally, of knowledge specific to the field of science, undeniably introduced a *radical rupture* with respect to other modes such as the mythical grasp of the world through language. It is therefore necessary to attempt to understand how the advent of such a rupture does not however preclude the differentiated unity of symbolic forms because if this rupture indeed introduces a new direction for sense, it does not however abolish the others which endure and which continue to have a reciprocal influence upon one another.

This double constraint implies considering that the philosophical problem of the place of language conceived as a foundation for symbolic forms in general *remains indefinitely open*: it is therefore not a matter of answering the question whether language plays or not a role of foundation for symbolic forms but to only remark that *is it because of the intrinsically mythifying nature of language that the foundational question poses itself for it*. If the issue of *foundation* is indeed linked to the *mythifying* nature of language, then a solution which would purely and simply elevate language to a status of foundation for symbolic forms would not enable to take into account the *ambivalent* place of language in the general project of a philosophy of symbolic forms, and would not enable to analyze how this ambivalence involves it in a *specific* manner in the relation it bears, each time, with other symbolic forms.[22] Such is the case, for instance, with the phenomenon of expression, as we will later see in more detail, which cannot limit itself to language[23] and which also extends beyond, even to the animal world[24] but which language transforms integrally to give it a profoundly original aspect in its own sphere of activity.

perception is totally lost or superfluous within the general view of reality which the theoretical consciousness projects."

[22]Cassirer makes similar remarks concerning law; E. Cassirer, *Axel Hägerström. Eine Studie zur Schwedischen Philosophie der Gegenwart*, ECW 21, p. 101: "It is impossible to derive the substance of law and the content of positive juridical propositions from any "original contract". But on the other hand, it must be granted that all juridical order require, in order to go on, a specific function that only the natural law tries to describe under the concept of contract."

[23]E. Cassirer, *The Philosophy of Symbolic Forms*, vol. 3, p. 213 (ECW 13, p. 245): "What Jackson has designated as the faculty of "statement", "the propositional" use of words, Head terms the faculty of symbolic expression and symbolic formulation. But he takes a significant step beyond Jackson in that he does not limit this symbolic function to language alone. To be sure, language is and remains the most evident exponent of this function, but language does not exhaust the entire range of its activities."

[24]E. Cassirer, *The Philosophy of Symbolic Forms*, vol. 3, p. 65 (ECW 13, p. 72) cites Köhler, "Zur Psychologie der Schimpansen", *Psychologishe Forschung*, I (1922): "There is a very great diversity of expressive movements, through which the animals "understand one another", although we cannot speak of any kind of language between them and cannot say that any particular movements or sounds possess a significative or representative function."

3.2 Symbolic Forms as Effects of Transcategorical Operations

Given what has just been stated, we will favor the second branch of the alternative of which we spoke above consisting in considering symbolic forms to be specific modes of expression worked by norms which do not have for finality the self-reflection of consciousness, but *sociality as such*. This line of thinking has, at least under a certain form, already been opened,[25] and it seems in our view to be more faithful to the evolution of Cassirer's thought after the publication, in 1923, of the first volume of *The Philosophy of Symbolic Forms*. We must indeed not lose sight, in order to progress regarding the delicate question of the role of language in the very architecture of a philosophy of symbolic forms, that it is less the notion of the diversity of forms than that of their *diversification* which is at stake and that this diversification does not seek to *file* already constituted objects under categories which have been set once and for all (this would leave us with a relation of subsumption and therefore with the classical distinction between the universal and the particular) but to *generate* (in accordance with Goethe's teaching) new semiotic fields constituting objects of knowledge or instituting modes of signifying by means of operators which are not limited to categories that were set beforehand over fields of objects.[26]

It is then a matter of thinking about symbolic forms as being effects of *transcategorical* operations ensuring in a same movement the transformation of significations and the transformation of the interpretive frameworks which make them possible,[27] that is, the joint transformation of signification and sense in the form of persistencies and ruptures.[28] If the effects of these transcategorical operations are found both in the

---

[25] I am borrowing here for my own purposes, even if it is for different reasons and especially in view of different objectives, the solution proposed by John M. Krois in "The priority of "symbolism" over language in Cassirer's philosophy", *Synthese*, 2008, where he speaks of a "symbolic turn" which is not specifically linguistic and which highlights even more primitive forms of semiosis, in particular those of technological and ritual activity. J. M. Krois concludes that Cassirer had realized as early as 1925 that the order of the three volumes of *The Philosophy of Symbolic Forms* as he had begun writing them in 1923 was inadequate and that he should have begun his book with mythical thought.

[26] In "Form and Technology", the distinction which Cassirer adheres to is that which exists between *forma formans* and *forma formata*; E. Cassirer, "Form and Technology", p. 18 (ECW 17, p. 142): "The world of technology [...] begins to open up and to divulge its secret only if we return from the forma formata to the forma formans, from that which has become to the very principle of becoming."

[27] E. Cassirer, *Axel Hägeström, eine Studie zur schwedischen Philosophie der Gegenwart*, ECW 21, pp. 86–87: "But here too the question is raised whether the insight into the genetic beginnings of culture can give us an insight into its various functions. [...]; the question is not what these functions whether language, art, law were originally but what they became by virtue of this transformation of meaning."

[28] One could be tempted to use the classical term "syncategorematic" instead of "transcategorical" as an adjective specifying the meaning of the notion of "operator". However, the almost exclusively logical usage of the term "syncategorematic" today could introduce ambiguity which the adjective "transcategorical" enables on the other hand to avoid.

types of activities as well in the modes of knowledge,[29] it is then necessary to conceive of the operations in question as *primitive phases, that are specific and unstable, in the process of the differentiation of forms*, and which are prior to the distinction between activity and knowledge.[30] These primitive phases must not be related in an essential manner to the self-revelation of a transcendental consciousness but to the constantly renewed reworking of modes of expression specific to the symbolic forms. It may then be asserted that the evolution of Cassirer's thinking leads to consider the symbolic forms, be they types of activities or modes of knowledge, and *regardless of their actual or future number*, as the result of a *play between three types of transcategorical operations* of which Cassirer has indeed given paradigmatic examples when he described the notions of language, myth, and science at some point of his intellectual career but which are not limited to them, as he realized himself as early as 1925 and as demonstrated in the remainder of his research pertaining to, among other things, art and technology. The effects of these three transcategorical operations can thenceforth no longer only be named "language", "myth", and "science" when a more generic designation is required. Continuing along this path,[31] we will attempt to interpret Cassirer's progression based on the way in which he presents what he calls the "phenomenology of knowledge" in the third volume of *The Philosophy of Symbolic Forms* since it is indeed starting with the publication of this volume, in 1929, that Cassirer modified once and for all the order of presentation of the forms and that he distinguished three moments, those of "expression", "representation",

---

[29]It is in this respect that I would diverge from Krois' interpretation which seems to consider activity and knowledge as two radically distinct spheres and, on the other hand, seems to consider that which pertains to activity (ritual and technical) to be *more primitive* than that which pertains to knowledge. The interpretation defended here tends on the contrary to develop the idea that transcategorical operations are *from the onset* assignable to the two dimensions of activity and knowledge in the production of all *forms*.

[30]It is in this manner, in our view, that the notion of "basis phenomenon" which Cassirer exploits in the later part of his life should be understood. Cf. J. M. Krois, "Symbolisme et phénomène de base (*Basisphänomene*)", *Revue Germanique Internationale*, 15/2012, p. 165: "Contrary to structuralist conceptions of symbolism, Cassirer's theory of Symbolic Forms was founded on a triadic notion of the symbolic process more than on the static and dual relation between signifier and signified."

[31]We follow here the interpretation by J. M. Krois who has shown that volume 3 of *The Philosophy of Symbolic Forms* comes to propose the new ordering of forms thenceforth beginning with the mythical form, but we would add, however, that this first form pertains rather to *language and myth*; cf. John Krois, "The priority of "symbolism" over language in Cassirer's philosophy", *Synthese*, 2008, p. 12: "Given this pragmatic reorientation, it must have been apparent to Cassirer by 1925 that the order of the first two books in The Philosophy of Symbolic Forms was incorrect. But it was too late, for the first volume had been in print since 1923. Indeed, when he wrote the third volume—the most important single book he ever published—expressive symbolism, the basis of mythical thought, came first, followed by the propositional type of meaning he called the "Darstellungsfunktion" (the function of representation)."

and "signification", without relating them to *objects* specific to particular forms such as language, myth, or science.[32] There is therefore a *radicalizing of the critical perspective* here in that it renounces keeping solely with objects, but returns to their conditions of possibility, the objects finally appearing as modalities of the *articulation* of these three moments.[33] The order of "expression", "representation", and "signification" will always remain with him from that point onwards, since it can be found elsewhere,[34] even if the formulation is modified somewhat, and it can even be found in his 1940 text on "Basis Phenomena" in which he attempts to grasp the specifically semiotic character of symbolic forms by using the three terms which he borrows from Bühler's *Theory of Language* (1934): "expression", "evocation", and "representation", terms which he relates at the same time to *Maxims and Reflections* (#391 to #393).[35] Other formulations can be found, for example when Cassirer remarks[36]:

> It is concrete thought, that is, the content of all cognitive acts in general, the content of all that leads to the "positing" of something objective (not only a sphere of the I and the You, but also an It-Sphere) and for which it is the indispensable condition.

We shall therefore call the three transcategorical operators "expression", "evocation" and "objectification", even if it is not exactly Cassirer's own wording and attempt to interpret the three terms as designating not only operations specific to language as was the case with Bühler, but also the transcategorical operations we have just mentioned. We shall therefore examine how their interplay enables to bring to light four of the symbolic forms studied by Cassirer: language, myth, science, and technology, the study of the latter form being only indirectly present in the second volume of *The Philosophy of Symbolic Forms* under the aspect of the specific technology of ritual, whereas it is directly thematized in his later work.

---

[32] It can be noted that it is on the basis of *knowledge* that the itinerary of forms which are not knowledge as such can be reinterpreted: Cassirer does not depart, in this respect, from a philosophy where epistemology plays a capital role.

[33] It is also for this reason that Cassirer directs his attention towards the relations between symbolic forms, in particular the relation between language and myth, which he studies in its own right (cf. E. Cassirer, *Language and Myth: A Contribution to the Problem of the Names of Gods* in E. Cassirer, *The Warburg Years (1919–1933)*, pp. 130–213 (ECW 16, pp. 227–311).

[34] For example, in "The Problem of the Symbol and Its Place in the System of Philosophy" in *E. Cassirer, The Warburg Years (1919–1933)*, pp. 254–271 (ECW 17, p. 261).

[35] Ernst Cassirer, *The Philosophy of Symbolic Forms*, vol. 4, p. 152 (ECN 1, p. 148): "Another aspect of the results of Bühler's analysis is still more significant for our observations as a whole. For what does this division of "language" into three basic aspects—expression, evocation, representation—signify, and to what does it finally direct us? If we consider it more closely, then we find to our surprise that it refers to the three classes of basis phenomena that we distinguished before, and that, for example, occur in Goethe's outlook". The maxims of Goethe are reproduced by Cassirer p. 127. It is to be noticed that Bühler terminology is even more complex than what Cassirer supposes since « representation » is the English translation of at least three German terms in Bühler's works: 'Vorstellung' (mental representation), 'Darstellung' (symbolic representation) and 'Vertretung' (proxy representation).

[36] Ernst Cassirer, *The Philosophy of Symbolic Forms*, vol. 4, p. 153 (ECN 1, p. 123).

Thus, the interpretation consisting in studying the interplay between these transcategorical operations has three main goals: first, to account for the possibility of a *plasticity of categories* through the conjunct modification of the latter and the objects they render manifest, according to the teaching Cassirer derives from Einstein's relativity; Second, *to go beyond the still categorical primacy of activity over knowledge* and at the same time to renounce any foundation of an hypostasied antepredicative nature in a consciousness for which symbolic forms would be mere "reflections", "emanations", or "self-revelation"; Third, to consider the semiotic modes specific to symbolic forms as being *public from the onset* and as serving as loci for the exploration of any form, be it of a perceptual, linguistic, or conceptual origin.

Chapter 5
The Three Operators of Semiosis: Expression, Evocation, Objectification

Abstract This chapter is an application of the results gained so far. It investigates the relationships between the three semiotic operators the dynamics of which establishes stabilized Symbolic forms. Four of these forms are under closer scrutiny: Myth and Ritual, Language, Scientific knowledge and Technology. In the case of the first operator, that of Expression, one can show that its power is based on a blending of idiomatic meaning and performativity that Cassirer calls "symbolic pregnance". Although the power of Expression is usually recognized in Myth, Ritual and Language only, it is also at work in Science and in Technology which are both nurtured by symbolic pregnances in a more concealed way. The second operator, called Evocation, is concerned with the possible transfer of symbolic pregnances which are always locally rooted in perception but can expand to larger, less immediate, fields. In this respect, the construction of a mythical space and time as well as the mythical mode of numbering are conceived as some of the most basic elements which pervade the mythical field. In the linguistic domain, two modes of transfer can be described: either an inductive one which leads to a substantial conception of generality or a functional one in which generality derives from the exemplarity of singular cases. These two forms of generality induce two different views on signification that participate in the inner and outer construction of "Symbolic forms". In Science, the way Cassirer uses the concept of group of transformation is typically a way that allows for a wide range of transfers, from perceived forms (Gestalten) to various forms of knowledge both in Mathematics, Natural and Human sciences. Technology has its specific mode of transfer as well for technology does not only operate with tools but on tools, specifying thereby the conditions of their use and generating specific domains for their relevance. Objectification is the last semiotic operator studied by Cassirer, once the purely arbitrary dimension of signs is acquired and accepted. It presents itself as a paradox, that of a semiotic construction which should not be interpreted as the result of a construction but as objective nature. This is the case in Mythical thinking which describes a natural order that does not depend on the whim of individuals but is subject to laws and (divine) will. It is also the case in Language where idiomaticity should not be conceived as a form of empirical resistance to conceptual knowledge but as a first step towards the construction of possible objects. As for Science, it is the very concept of nature which presents itself in a problematic way since Science appears as the attempt to conceptually construct a science

© Springer Nature Switzerland AG 2020   121
J. Lassègue, *Cassirer's Transformation: From a Transcendental
to a Semiotic Philosophy of Forms*, Studies in Applied Philosophy, Epistemology
and Rational Ethics 55, https://doi.org/10.1007/978-3-030-42905-8_5

of nature that does not depend on scientific mediations. In the case of Technology, it becomes clear that the concept of nature produced by technological mediations questions the very independence of nature although technology itself only claims to copy a pre-existing nature.

Keywords Expression, evocation and objectification as trans-categorial operators · Expression as the most basic sense-making process · Evocation as transfer · Objectification as modes of objectivity · Case studies: trans-categorial operators in myth and ritual, language, scientific knowledge and technology

The relations of stability and rupture between symbolic forms can be explicated by the interplay of the transcategorical operators of which it has just been question and which Cassirer calls idiomatic expression, participative evocation, and objectification. It is by adopting the standpoint of this ternary interplay in the mediation operated by signs that we will study the examples provided by four particular symbolic forms pertaining to language, myth, scientific knowledge, and technology,[1] and we will also study their relations by basing ourselves on the way in which each form specifies the role played by the mediation of signs.

1   Expression

The notion of expression is *originary* in the sense that it situates itself prior to the division between activity and meaning, the signifying material being at first indistinct which only later will appear as the forms of language, myth, ritual, knowledge. It is by a *progressive drift* from this primary indistinction that symbolic forms pertaining more specifically to activity (such as ritual and technology) or more specifically to knowledge (such as myth or scientific knowledge) will finally be individuated, with language occupying a median position, a position which is quite particular and which is the fruit of intense reflection on the part of Cassirer and to which we will return.

1.1   Expression Inherent to Form Perception

In the broadest sense, associating "expression" to "form" perception amounts to considering that perceiving a form consists in attributing it to an *internal spontaneity conveying a value which distinguishes it from a background*. That a perceived form,

---

[1]It should be noted from the start that the analysis of technology in the works of Cassirer is mostly to be found in his 1930 article "Form and Technology" as well as in some dispersed remarks in *The Philosophy of Symbolic Forms*. The analysis of technology therefore does not at all have the same scope as do his reflections on language, myth, or science which have been an important and prolific focus throughout his whole life, hence the somewhat lopsided character of the remarks which will follow.

for instance a darker area of the sky, would be immediately conceived as a threat,[2] that a face would appear as friendly or as inimical,[3] that a sentence would manifest a tonality in the particular configuration of the discourse within which it takes place,[4] all of this is familiar to us without us generally seeing there anything else than contingent subjective projections. Now, the expressive character inherent to a form is not only subjective *because it is not added on top of it*.[5] It is, on the contrary, the manifestation of its internal spontaneity: operating a condensation around a pole constituted by configuration and value setting it apart from a background, whether it stems from the perception of the natural environment or from the institution of signs, it is indeed *addressed* to a recipient without constituting however a subjective or contingent point of view. Cassirer calls this relation of expressivity inherent to any perceptual act *symbolic pregnancy*: such an act is never decomposable into a neutral apprehension of physical contours to which would be added in a second phase a meaning projected from an intellect, as the intellectualist theories of perception of the seventeenth century would argue. As Gestalt theory, of which Cassirer had followed the developments, has abundantly shown, expressivity is generated at the level of *perception itself* and constitutes *the very modality by which a form distinguishes itself from a background*. Endowed with an *internal coherence* enabling to distinguish an inside from an outside and with a *plasticity* susceptible of preserving it throughout an evolution, be it spatio-temporal[6] or semantic, the form is therefore

---

[2] E. Cassirer, *The Philosophy of Symbolic Forms*, vol. 3, p. 66 (ECW 13, pp. 72–73) cites Köhler, "Komplextheorie und Gestalttheorie", *Psychologische Forschung*, I, 1922, p. 27: "May it not be peculiar to certain formations", Koehler continues, "that they essentially bear the character of the terrible, the forbidding, etc., not because an innate ad hoc mechanism enabled them to do so but because, in a given state of the psyche, certain Gestalt conditions necessarily and in accordance with an empirical law produce the character off the terrible, while others call forth that of the charming, the awkward or the energetic and severe?"

[3] E. Cassirer, *The Philosophy of Symbolic Forms*, vol. 3, p. 64 (ECW 13, pp. 71–72) cites Koffka, *Die Grundlagen der psychischen Entwicklung*, (1921): "The infant is not interested in simple colors, but in human faces… And as early as the middle of the first year of life the effect of the parents' facial expression on the child can be established. […]. And yet by the second month the child knows his mother's face; by the middle of his first year he reacts to a friendly or an angry face, and so differently that there is no doubt that what was given to him phenomenally was the friendly or angry face and not any distribution of light and dark."

[4] E. Cassirer, *The Philosophy of Symbolic Forms*, vol. 3, p. 449 (ECW 13, p. 521): "Language, it is true, discloses the new turn, the transition to a new dimension, more clearly than the other two; we cannot doubt its connection with the world of expression. There is always a certain expressive value, a certain "physiognomic" character in words, even in those of highly developed language."

[5] See in this respect Cassirer's unpublished text "Zur 'Objektivität der Ausdrucksfunktion'" cited in J. Krois, "Ernst Cassirer's philosophy of biology", *Sign Systems Studies* 32.1/2, 2004: 277–295.

[6] E. Cassirer, "The Concept of Symbolic Form in the Construction of the Human Sciences" in E. Cassirer, *The Warburg Years (1919–1933)*, p. 75 (ECW 16, p. 78): "It is a question of taking symbolic expression, that is, the expression of something 'spiritual' through sensory 'signs' nd 'images', in its most general signification; it is a question of asking whether this form of expression, with all its different possible applications, is grounded by a principles that marks it as a closed and unified fundamental process; thus, what should not be asked here is what the symbol signifies and achieves in any particular sphere, what it signifies or achieves in art, myth, or language, but how far language

not assimilable to a substance which would liken it to inert material but rather pertains to a deployment and interaction with other forms and an environment. From this point of view, the notion of expression is of a dynamic nature and indicates both the progression through which a form had been shaped and the end towards which it tends, within an environment which Cassirer generally calls the "background", by directly borrowing the concept from Gestalt theory which was the first to theorize this opposition.[7] This distinction thus enables to account both for the universality and for the specificity of forms in the dynamic relation they bear with the backgrounds from which they emerge but which can themselves become forms with respect to another background, the background/foreground opposition being, in some cases, reversible.[8] In the wide semantic field pertaining to the notion of form, this Gestaltist point of view allows to avoid a pitfall into which it is easy to fall when seeking to clarify the relation between natural and cultural forms.

1.2   Darwinism and Expression

Equipped with such a broad definition of expression, one might believe that the expressivity of forms enables, firstly, to establish a continuity between all forms, be they natural or cultural and, secondly, to relate cultural forms to natural forms considered to be the most primitive because they existed "before" the rise of cultural forms, even if the chronological anteriority in question remains generally non-specified and, in the end, thoroughly obscure. Cultural forms, such as language or myth, would then finally only be "sublimated" natural forms of which it would be possible to find the trace. Cassirer links the terms "sublimation" back to Darwin as presented in *The Expression of the Emotions in Man and Animals*: the "sublimation" specific to cultural forms would then result from an *inhibition of action*.[9] It

---

as a *whole*, myth as a *whole*, art as a *whole*, carry within them the general character of symbolic configuration."

[7] The difference is first made explicit by the Danish psychologist Rubin in 1921 before reappearing in the works produced by the Berlin school of Gestalt, in particular those by Werner, who is often cited by Cassirer. Cf. V. Rosenthal and Y.-M. Visetti, *Koehler*, (2003), p. 129 sq.

[8] E. Cassirer, "Form and Technology", p. 37 (ECW 17, p. 167): "Each new gestalt of the world opened up by these energies is likewise always an opening out of inner existence; it does not obscure this existence, but makes it visible from a new perspective. We always have before us a manifestation from the inner to the outer and from the outer to the inner - and in this double movement, in this particular oscillation, the contours of the inner and the outer world and their two-sided borders are determined."

[9] E. Cassirer, *The Philosophy of Symbolic Forms*, vol. 1, p. 180 (ECW 11, p. 125): "In his work on *The Expression of the Emotions in Man and Animals* Darwin attempted to create a biological theory of expressive movements by interpreting them as a vestige of actions which originally served a practical purpose. According to this theory, the expression of a specific emotion would be merely an attenuated form of a previous purposive action: the expression of anger, for example, would be merely a pale, attenuated image of a former movement of aggression, the expression of fear would be the image of a movement of defense, etc."

is from this inhibition that cultural forms would derive their status as *signs* referring to actions which would have *not been carried through*, thereby keeping in this respect with the classical definition of the sign: "something that stands for something else". Then, a clear separation would establish itself between natural forms, the only ones designated as "real", and cultural forms, which would lie "in their stead". This separation would amount to attributing no autonomy to cultural forms all the while entrenching their status as signs as being mere reflections: the cultural forms would only be simple copies of natural originals, and one may wonder why and how they came to be because the reason or reasons for the inhibition of action governing the manifestation of their expression is generally not further explicitated. The attractiveness of this type of conception stems from the fact that it seems possible to easily establish continuity *but especially a hierarchy* between "nature" and "culture" by resolving both the problem of the specificity belonging to each sphere as well as the issue of their reciprocal relationships.

Cassirer criticizes this point of view and proposes to re-elaborate yet another Darwinian concept, that of *mutation*, a concept which challenges that which the notion of inhibition had implicated with respect to the definition and role of expressivity: a form can indeed appear in continuity with other forms—be they biological or cultural—but this is however nothing but appearances because it is the fruit of a *rupture* having completely disrupted the relation it had with the background by instituting *another type of values to promote or to pursue*. We may think for example of the familiar case of mutations in living species: we indeed see that the appearance of a new species institutes a new relation to the environment in the one the species bears with the nutritional and vital resources (such as the aquatic, terrestrial, or marine elements), its relations with congeners and with other species, etc. But we could also invoke cultural examples of a technological nature: for Cassirer, the sewing machine introduced a new way of sewing, while the steel mill introduced a new metallurgy.[10] It is therefore not the continuity between forms and particularly not the "transition from the natural to the cultural" which must be sought to be updated, because this "transition" itself is purely imaginary. A mutation is not a "transition", this being an expression which leads to think that it is always the same values and the same categorical framework which are to be found on each side of a turning point in the progressive evolution of a form, one which is finally identical with itself regardless of circumstances.[11] A mutation is indeed a *rupture* in the sense where, as the advent of a new form, it requires an integral reconfiguration of its relation to the background.

---

[10]E. Cassirer, "Form and Technology", p. 39 (ECW 17, p. 169): "The discovery of new tools represents a transformation, a revolution of the previous types of efficacy and the mode of work itself. Thus, as other thinkers have emphasized, with the advent of the sewing machine comes a new way of sewing, with the steel mill a new way of smithing – witness the problem of flight, which could be only finally be solved once technological thinking freed itself from the model of bird flight and abandoned the principle of the moving wing."

[11]E. Cassirer, "The problem of Form and the Problem of Cause" in *The Logic of the Humanities*, Yale University Press, 1960, pp. 179–180 (ECW 24, p. 460): "Today, the Darwinian theory, which once sought to carry through this interpretation because of its predilection for the principle of continuity, is probably no longer defended by any biologist in the form which dogmatic Darwin gave to it. As a result, the thesis, "*natura non facit saltus*", underwent a most essential qualification. Its problematic

This reconfiguration abolishes thereby its own past without however replaying an absolute beginning, an "origin" which would go from a radical absence of form to its presence.[12] We know that Darwin had begun to break with the idea that evolution would tend towards an end and that a biological form would thereby be the result of an *optimal* adaptation to this end given its environment.[13] Optimality would indeed require the objective stability of a reference framework for given values, a framework which precisely lacks if we reject all recourses to teleological principles. Here also, all forms deploy their own values in the idiosyncratic relation they bear with their background without these values however being necessarily maintained when a mutation intervenes. In what concerns the debate between nature and culture, there is therefore neither a simple continuity between all forms nor a hierarchy which would favor natural forms with respect to their cultural "copies", once the latter have appeared. As Cassirer pointed out, there is indeed no external point of view which would enable to sort out any hierarchy between natural and cultural forms: the most elementary of perceptions shows this, as it immediately interprets a spatio-temporal configuration as the manifestation of a value, contrary to all intellectualist theories of perception which consider that the detection of configurations constitutes the first and fundamental stage of perception, the projection of value over these configurations coming only in a second stage[14]:

> Rather, it is the perception itself which by virtue of its own immanent organization, takes on a kind of spiritual articulation – which, being ordered in itself, also belongs to a determinate order of meaning. In its full activity, its living totality, it is at the same time a life "in" meaning. It is not only subsequently received into this sphere but is, one might say, born into it.

---

nature has been disclosed in quantum theory in the filed of physics and in mutation theory in the filed of biology. For in the organic sphere "evolution" remains nothing but an empty word if we are to assume that what is involved is a mere "unfolding" of a given and finished being, as in the old theories of preformation and involution, where in the end, every new form is enclosed within the old. Here, too, at each point we must acknowledge something novel, something not arrived at without a "gap"."

[12] E. Cassirer, *The Philosophy of Symbolic Forms*, vol. 4, p. 38 (ECN 1, pp. 36–37): "We can never penetrate back to the point at which the first ray of intellectual consciousness broke out of the world of life; we cannot put our finger on the place at which language or myth, art or knowledge "arose". For we know them all only as something already existing, as closed forms in which each particular carries the whole and is carried by it, and in which we therefore cannot indicate what is "earlier" or "later", temporally "first" or "second". [...]. Every analysis that assumes it is possible to "explain" this whole by dissolving it, so to speak, into its atoms of meaning thereby commits a fundamental error. The very concept of an aatom of meaning contains an internal contradiction."

[13] For Cassirer, the divorce from final causes, even if it began with Darwin, is not, in his case, completely achieved and he sides in this respect with the opinion of L. von Bertalanffy whose *Theoretische biologie* he cites extensively, in which the latter criticizes the Darwinian concept of adaptation; cf. E. Cassirer, *The Logic of the Humanities*, "The Problem of Form and the Problem of Causality" p. 169 (ECW 24, p. 453): "Actually, in its absolute formulation, the holistic mode of study has been as abused as was the "study of purposiveness" in its efforts to discover utility-values and selection-values for every organ and characteristic [...]." Cf. also J. M. Krois, "Ernst Cassirer's philosophy of biology", *Sign Systems Studies* 32.1/2, 2004: 284.

[14] E. Cassirer, *The Philosophy of Symbolic Forms*, vol. 3, p. 202 (ECW 13, p. 231).

This is the reason why it is necessary to conceive of mutations between forms not as the result of simple inhibitions of action, but as the always increasing *investment* of action in forms *having not only vital utility for a theme,*[15] *but also symbolic values.*[16] This change in perspective displaces the "Darwinian" scission between natural and cultural forms and restitutes it within the progressive differentiation of forms within a global process which has for origin perception, which is *semiotic from the onset.* Due to this, the status of the sign has changed: it is no longer entrenched within a particular sphere—that of "culture" as opposed to "nature"—where it would only play the role of a simple reflection of an action which would precede it, but it is on the contrary henceforth conceived as *present everywhere forms are found,* at the very core of the process of their development.

1.3   Adherence and Distance with Respect to Expression

No form is therefore devoid of expressive value since expressivity is consubstantial with form, but it is the attitude and method to adopt with respect to this expressivity which is likely to change and which institutes new forms. Cassirer notes that it is necessary to suppose a state in which what will appear, later on, as symbolic forms decomposed into language, myth, or even science begins by being relatively indistinct. There is thus a primitive mode of objectification which is at the same time linguistic, mythical, ritual, and gnoseological which *is primordially real* and with respect to which symbolic forms individuate themselves progressively. It is the relation of adherence or of distance with respect to this primitive expressivity which generates a diversity of forms of which the objectification modes become relatively autonomous with respect to one another. This progressive dissociation defines an asymmetry between forms, an asymmetry which goes from the less distinct to the most distinct and which constructs, thereby, a *direction through time.* Such is the case, for example, with language and myth of which the relative reciprocal adherence nevertheless finishes by authorizing an attitude generating on the contrary a

---

[15]E. Cassirer, *The Philosophy of Symbolic Forms*, vol. 4, p. 41 (ECN 1, p. 39): "The theoretical element of "seeing" did not result from a decline in the element of action; rather, it was the other way around: action separated more and more from its original basis in life, from its merely "vital" direction, as it became mixed with intellectual forms." Even if this unpublished text has a "Platonic" aspect in the way it appears to defend the possibility for disembodied intellectual forms which would progressively become accessible—which fits poorly with Cassirer's project as such—it nevertheless has the advantage of shifting the focus from the idea of *inhibition* of action towards that of *investment* of action in other directions than those of "life" as such. The symbolic forms are the result of these investments in directions which are both diverse and non-definite.

[16]This is the case for example with technology. Far from developing only a utilitarian relation to the world, technology is *symbolically* invested in reality; cf. E. Cassirer, *The Philosophy of Symbolic Forms*, vol. 4, p. 41 (ECN 1, p. 40): "Tools can arise only where the mind has become capable of conceiving of a "possible" object instead of giving itself over directly to a real one and losing itself in it. A consciousness of this new attitude and reality is expressed in the fact that tools are not only created and made use of, but they are also worshiped."

certain defiance with respect to the one or the other: it is this defiance which provoked, according to Cassirer, the advent of philosophy and of science during a *crisis*
in Greece during the sixth to fifth centuries B.C.,[17] a crisis which instituted a foundational divorce between the world of language and that of myth on the one hand,
and between the world of science and that of philosophy on the other hand.[18] It is
therefore both possible for Cassirer to think of the common birth of philosophy and
of science as being the product of a skepticism regarding the "mythifying" powers of
language *and* to reintegrate language and myth into the orbit of philosophy by means
of a philosophy of symbolic forms. How should this double constraint—adherence
and detachment[19]—be envisioned as it manifests throughout the course of the progressive exfoliation of symbolic forms starting with a primordial undifferentiated
signifying form? The notion of crisis poses, in this respect, two types of difficulties.

1.3.1   Adherence and Detachment Regarding the Expressivity of All Forms

The first consists in remarking that if there is a *crisis*, it is because it is possible to
go beyond the adherence of expression to the form, as the individuation of the form
"science" has shown in the case of Ancient Greece when it separated from language
and myth. Now, if it is indeed through expressivity that any form constitutes itself, it
appears difficult to conceive that this expressivity may disappear without at the same

---

[17]E. Cassirer, *The Philosophy of Symbolic Forms*, vol. 3, pp. 16–17 (ECW 13, p. 18): "The peculiar
character of philosophical concepts and the historical conditions of their development make it quite
understandable that philosophy should have been relatively late in taking up the totality of the
problems of form involved in myth and language, that it should long have avoided and even rejected
these problems. For the concept of philosophy attains its full power and purity only where the world
view expressed in linguistic and mythical concepts is abandoned, where it is in principle overcome.
[…]. And natural science arrives at the mastery of its specific task in very much the same way as
pure philosophy. In order to find itself it, too, must first effect the great intellectual differentiation,
the *krisis*, separating it from myth and language."

[18]E. Cassirer, *The Philosophy of Symbolic Forms*, vol. 3, pp. 16–17 (ECW 13, p. 18): "This act of
separation marks philosophy's hour of birth, and also the starting point of empirical research and
the mathematical determination of nature. In the beginnings of Greek philosophy the two problems
are still one and the same."

[19]It is this difficulty which, according to Cassirer, Oswald Spengler failed to resolve in his philosophy
of history: By limiting any manifestation of the form to expression alone, Spengler neglects what
is likely to depart from it, all the while purporting however the usage of analogies and even of
calculus, a usage which would enable, according to Spengler, to anticipate the historical events
to come and would thereby completely exceed the immediate aspect of expression. In Cassirer's
view, this contradiction stems from the fact that Spengler failed to see that the problem of the
history of forms arises when their becoming is to be understood on the basis of the *couple* formed
by the adherence to expression and the possibility for a detachment from it: "This is because for
Spengler the entire achievement of the symbolic function is limited to the function of expression."
And Cassirer concludes in *The Philosophy of Symbolic Forms*, vol. 4, p. 109 (ECN 1, p. 106):
"A consciousness that consisted simply of expressive values and that was able only to understand
purely expressive phenomena would in fact be as little able to think in terms of analogies as it would
be able to think in terms of mathematical laws."

time causing the form itself to disappear. This therefore means that it is necessary to conceive of this rupture in a way that it will never be without adherence with respect to the form: the crisis does not entail a pure and simple divorce in the relation between originative expressivity and form. There are *various modalities* of the adherence of expressivity to the form and it is precisely this point which enables to identify forms as particular modes of adherence and of detachment with respect to expression: it is in particular what distinguishes the forms of scientific knowledge and of philosophical understanding *from all others*, as we shall now see.

Cassirer insists on many occasions upon the primary indistinction of language and of myth from the standpoint of expression[20] and, thereby, everything, at first, seems to oppose myth and language on the one hand—in which the process of accumulation of knowledge is always put into question due to the lack of access to resources garnering innovations and transformations, in short, due to the absence of *recording*—and scientific knowledge on the other hand—in which, on the contrary, innovations and changes in perspective are duly recorded within a tradition and can always be subject to re-elaboration. However, it is necessary to remark that this frontal opposition between two great modes of objectification must itself by mitigated because one must acknowledge that scientific knowledge *itself* comprises a form of expressivity when it is possible to consider it with sufficient historical distance: as we have already seen for example in a previous chapter, Cassirer points out that Aristotelian logic is unconsciously linked to an expressivity specific to a certain historical state of the Greek language. More generally, Cassirer has shown as early as 1906 in his book *Substance and Function* that the very notion of *substance* as opposed to that of *function* should be interpreted as the primitive form of expressivity in the order of scientific knowledge, an expressivity of which we retrospectively see the unconscious influence after the category of function has been historically acquired and that its mastery has spurred us into the so-called "modern" period.

From this point of view, language and myth distinguish themselves from scientific knowledge due to the fact that the latter is endowed with its own dynamic which neither language nor myth are capable of developing and which consists in *distancing itself* from expressivity without however breaking with it[21]: knowledge becomes scientific when its mode of objectification tends to conceive of the existence of a "nature" which is *neutral of any expressivity*, this mode of objectification only becoming relevant through the *controlled* introduction of theoretical principles which filter the mythical, linguistic or ritual aspects all the while managing to account for the appearance of regularities without attributing them to one or several forces materialized in

---

[20]This is the object of his book Language and Myth: A Contribution to the Problem of the Names of Gods (ECW 16, pp. 227–311).

[21]E. Cassirer, *Language and Myth: A Contribution to the Problem of the Names of the Gods, in* Ernst Cassirer, The Warburg Years (1919–1933), pp. 137–138 (ECW 16, p. 79): "Mythical apprehension and interpretation are not subsequently introduced into certain *elements* of empirical existence; rather, the primary "experience" itself is thoroughly penetrated by the figures of myth and, as it were, saturated by its atmosphere. The human being lives with *things* only because and insofar as he lives in these *figures*."

volitions.[22] Thus, scientific causality, for example, consists in extracting from the mythical notion of volition—be it beneficial or malevolent—a schema which preserves its intelligible aspect without falling into either ad hoc explanations or into paranoia.

1.3.2   The Possibility for a Conscious Modification of the Modes of Objectification

The second difficulty concerning the notion of crisis comes from the fact that if there is a crisis, it is because it is possible, in the case of culture, to *consciously* modify the relation to the most primordial objectification mode, that of the form which is at the same time mythical, linguistic, and gnoseological. This conscious modification of the mode of objectification is specific to scientific knowledge and to philosophy which, without being a distinct symbolic form since it does not have its own terrain, is nevertheless characterized as a reflection regarding the possibility of modifying the modes of objectification in all symbolic forms. That it may be possible to consider the common birth of philosophy and of science to be the product of a crisis of defiance regarding "mythifying" powers of language and of ritual, while it may also be possible to reintegrate language, myth, and ritual into the orbit of philosophy through a philosophy of symbolic forms as was Cassirer's project, shows the both original and contemporary kinship between science and philosophy since the latter is also capable of modifying its relation to its own mode of objectification by reintegrating, over the course of its history, that which it began by casting aside. From this point of view, the possibility for such modifications in modes of objectification seeks not only to attempt to *separate* expressivity from forms, as is particularly the case regarding the form which is scientific knowledge, but also to *reintegrate* the very notion of expressivity into philosophy's own area of competence. We then understand that Cassirer may have for philosophical project the reintegration of myth and language into the sphere of philosophical rationality, instead of maintaining them separate, as was the case at the moment of the advent of science and philosophy in Ancient Greece. But then, a question arises. Cassirer described in his posthumous book *The Myth of the State* the stranglehold the Nazis exercised over German culture as being the result of a conscious manipulation aiming, in particular by misappropriating the signification of words and by reviving emotions linked to ancient Germanic myths, to unconsciously constrain subjects and to enroll them despite themselves into a totalitarian society. In what does the *conscious manipulation* of language, myth, and ritual using appropriated technologies such as was practiced by the Nazis distinguish itself from

---

[22] E. Cassirer, *The Myth of the State,* ECW 25, p. 54: "The beginning is not simply a commencement in time but a "first principle"; it is logical rather than chronological. According to Thales, the world *was* not only water, it *is* water. Water is the abiding and permanent element of all things. From the element of water or air, from the "Apeiron" of Anaximander, things have developed not in a haphazard way, according to the whims and caprices of supernatural agents, but in a regular order and according to general rules. The concept of such inalterable and inviolable rules is perfectly strange to mythical thought."

the *conscious reintegration* of language, myth, and ritual into the sphere of a philosophy of symbolic forms? This is a legitimate question given that both operations seek to *consciously rework* the types of objectification specific to the same forms. The two operations, however, do distinguish themselves from one another, and not only in their radically divergent ethics. In the case of philosophy, this reintegration is carried out *under the aegis of theory* in view of an act of greater *understanding* and not in order to arouse emotions and to direct them towards ends which do not pertain to understanding. This leads us to a better understanding of the kinship between science and philosophy: without this original *theoretical* kinship, the reintegration of forms as primitive as language, myth, and ritual into the sphere of rationality and of consciousness would turn into pure and simple regression for those who would fall victim and would provide all sorts of opportunities to those who would seek to exercise their power. From this perspective, the purely theoretical character of the kinship between science and philosophy serves as a safeguard enabling scientific and philosophical rationality to *deploy a direction over time* by founding a history taking the form of a progressive accumulation and always enabling accessibility at each of its stages. There is therefore no objective direction in the deployment of forms which would transition through mandatory stages but only a direction which humanity gives itself and which aims for the *general* increase of positive knowledge under the aegis of productive knowledge and not by the manipulation of the ones by the others. It is therefore, in Cassirer's view, by keeping with the act of understanding that it is possible to avoid the pernicious effects of the reintegration of the strata of meaning belonging to the most primitive of forms, such effects which would have for consequence to undermine the progressive construction of a history endowed with a direction.

It is now possible to more fully delve into the analysis of the notion of expression in the four symbolic forms which will serve here as examples: myth, language, science, and technology.

1.4 Idiomatic Expression in Myths

It is impossible to not be stricken by the contradictory, chaotic, and, to use a single word, phantasmagorical character of mythical narratives: the origin of the cosmos being attributed to combats between giants, heroes returning from the world of the dead, protective, vengeful, or lecherous gods transforming their physical appearance in order to carry out their egoistic deeds towards men and women, talking animals, plants, or minerals endowed with magical powers, metamorphoses of all sorts, prophets who can communicate with angels or gods, individual or collective salvation dependent upon trials of which the ultimate meaning remains impenetrable, lacking and incoherent narratives—these prodigies, as varied as they are obscure, make myths as enigmatic to those who believe in them as to those who do not or no longer do. By placing himself within the tradition inaugurated by Schelling who already attributed an intrinsic value to myth and who no longer interpreted myth in an allegorical manner but rather as a creation of signification endowed with its own autonomy,[23] Cassirer too broke with the quest of uncovering "objective" mythical elements which would be considered, for more or less contingent reasons, to be the most fundamental—the forces of nature, spiritual powers, or natural catastrophes—and which would provide the elements for a general theory of the *hidden* signification of myths. By placing himself from the point of view of the analysis of objectification modes, Cassirer sought to describe the existence of a mythical "function" of *thought in general*[24] which would take into account the occult and lacking character of mythical elements without this character depending on the objective content of these elements or on their transmission, but rather on the very particular function that sense plays for them. And indeed, if we were to keep with the interpretation according to which it is the objective content of the mythical elements which can account for their phantasmagorical aspect, it would be necessary to give the reasons for which the signification of mythical elements would have ended up being occulted over the course of time, which would entail the risk of uncontrollably projecting human history onto a brand of "primitivism" operating a scission between a primordial state (one which would be real but now hidden) and a second state (diminished and manifest) occulting the first state, quite analogously to the way in which Darwin, according to Cassirer, distinguished nature from culture. It is, for Cassirer and on the contrary, a matter of *keeping with the mythical function of sense as it is given, in other words, with its intrinsic signification.*[25] Although Cassirer does not explicitly thematize this point, keeping with meaning as it presents itself implies considering that *obscurity is constitutive of the expression of sense in myths* and not only a hazard of its transmission. Sense expresses itself in myth in this particular light, which is to be respected as such if we wish to account for its specific nature: there is therefore no reason to *decode* the obscurity of myth in an attempt to reduce it, thereby uncovering a clarified concept of its nature but, on the contrary, there is

---

[23] E. Cassirer, *The Philosophy of Symbolic Forms*, vol. 3, p. 62 (ECW 13, p. 68): "Consequently, if the philosophy of mythology is to meet the fundamental demand first made by Schelling, if it is to understand myth not only allegorically as a kind of primitive physics or history, but categorically as a symbol of independent significance and form, it must do justice to that form of *perceptive experience* in which myth is originally rooted and from which it forever draws new nourishment."

[24] E. Cassirer, *The Philosophy of Symbolic Forms*, vol. 2, p. (ECW 13, p. 25): "In the last analysis this unity must be established not in agenetic and causal but in a *teleological* sense – as a direction followed by consciousness in constructing spiritual reality. Regardless of whether we gain an understanding of its genesis and regardless of what we view we take of this genesis, the reality that is produced in the end stands before us as a self-contained configuration with a being and a meaning of its own. And myth, although it is limited to no particular class of things or events but emcompasses the whole of existence, and although it employs the most diverse spiritual potencies as its organs, represents a unitary perspective of consciousness from which both nature and soul, both "outward" and "inward" being, appear in a new form."

[25] E. Cassirer, *The Philosophy of Symbolic Forms*, vol. 2, pp. 37–38 (ECW 12, p. 46): "We are accustomed to view these contents as "symbolic", to seek behind them another, hidden sense to which they mediately refer. Thus, myth becomes mystery: its true significance and depth lie not in what its configurations reveal but in what they conceal. [...]. From this result the various types and trends of myth interpretation – the attempts to disclose the meaning, whether metaphysical or ethical, that is concealed in myth."

reason to take heed of this obscurity, even as manifested at any moment in time (and not through a projection over the past, this already being of the order of myth).

This enables to better understand that which distinguishes the form of myth from that of science: the latter consists precisely in *making intelligible* this constitutive obscurity by a introducing a distance between expressions and forms that are given as united in the case of myths. We also understand that the signifying material specific to myth would then be susceptible of diffracting into autonomous forms depending on whether expression in myth is fully adherent to the form or, on the contrary, tends to be neutralized in myth.

1.4.1 Emotional Intensity as Collective Expression

The mythical form is therefore not a hidden signification but indeed a *specific expression* of sense which consists above all in the intensification of a sensible impression,[26] itself being emotionally charged[27]: then, a particular perspective on sensible reality constitutes itself—a "pregnancy"—which confers to this point of view a status of method organizing the chaotic flux of impressions into a perception having emotion for principle of coherence.[28] Mythical signification then takes the aspect of a world where natural forces manifest as *will* regulated by finalities.[29] The sensible elements underscored in emotion objectivate, or maybe it would be preferable to say that they

---

[26]E. Cassirer, *The Philosophy of Symbolic Forms*, vol. 2, p. 35 (ECW 12, p. 44): "Myth lives entirely by the presence of its object – by the intensity with which it seizes and takes possession of consciousness in a specific moment."

[27]E. Cassirer, *An Essay on Man*, ECW 23, pp. 90–91: "His [Primitive man] view of nature is neither merely theoretical nor merely practical; it is *sympathetic*. If we miss this point we cannot find the approach to the mythical world. The most fundamental feature of myth is not a special direction of thought or a special direction of human imagination. Myth is an offspring of emotion and its emotional background imbues all its productions with its own specific color. Primitive man by no means lacks the ability to grasp the empirical differences of things. But in his conception of nature and life all these differences are obliterated by a stronger feeling: the deep conviction of a fundamental and indelible *solidarity of life* that bridges over the multiplicity and variety of its single forms."

[28]E. Cassirer, *Language and Myth: A Contribution to the Problem of the Names of Gods*, p. 166 (ECW 16, p. 264): "Here too, as it appears, the direction of this progress is determined primarily through the direction of doing something; what is reflected in the form of mythical configuration is not so much the objective form of things as the form of human effective activity."

[29]E. Cassirer, *The Philosophy of Symbolic Forms*, vol. 2, p. 49 (ECW 12, pp. 67–68): "Myth, however, takes the opposite path. It begins with the intuition of purposive action – for all the forces of nature are for myth nothing other than expressions of a demonic or divine will. This principles constitutes the source of light, which for myth progressively illuminates the whole of reality and outside of which there is no possibility of understanding the world."

dramatize this intensity in magical or divine powers which Cassirer calls "demonic".[30] For Cassirer, ritual[31] directly pertains to this emotional dramatization which it interprets as both a collective and bodily expression. Cassirer takes the case of war in archaic societies as an example: while the men go to battle, the women dance, their dance aiming to produce victory.[32] The signification of the ritual manifests through the rhythm imprinted upon the bodies of the dancers in accordance with the collective rules surrounding the particular dance which seeks a collective goal, the victory of the group obtained by means of the battling men: the effectiveness of the dance is therefore not causal at all, in the sense that scientific knowledge gives to this word, but it is rather *participative*, thereby implying both an emotional and collective goal.[33] The collective transaction which takes place through participation supposes that the element which materializes the collective interest—in our example, the dance—would be endowed with a form that is *public to begin with* and that has a certain modality[34]—the *way* in which the dance is performed—through which may be embodied the *collective* emotion of actors reinforcing their collective finalities, esthetic values, and ethical norms by means of the dance itself. From this perspective, emotion is not limited to the physiological aspect of individual bodies: its symbolic charge, when it is inscribed within a ritual, is not attenuated after the bodies tire but preserves an intensity which is precisely linked to its symbolic aspect.[35]

---

[30]E. Cassirer, *The Philosophy of Symbolic Forms*, vol. 4, p. 70 (ECN 1, pp. 67–68): "The world of expression as such is nowhere transcended, but a concentration takes place in it, a gathering together of particular "points of judgement". Only in this gathering together and combination does it become the *demonic* world. The demon may be conceived as ever so vague and fleeting, but it always has some kind of personal "character" by means of which it can be distinguished and recognized. It is helpful or hostile, cruel or good, ready to give protection or malicious or treacherous. It maybe be quite unpredictable in its behavior, moody and changeable in its particular expressions, yet it possesses certain limits to its essence and particular character from which it does not depart.[...]. So here too in these first, primitive structures of myth the decisive step has already been made. The chaos of affects has begun to clear, and particular configurations emerge from it that acquire an enduring nature."

[31]E. Cassirer, *The Myth of the State,* ECW 25, p. 31: "Myth is the *epic* element in primitive religious life; rite is the *dramatic* element."

[32]E. Cassirer, *The Myth of the State,* ECW 25, p. 40: "If in a savage tribe the men are engaged in warfare or in any other dangerous enterprise and the women who have stayed at home try to help them by their ritual dances – this seems to be absurd and unintelligible when judged according to our standard of empirical thought and "causal laws"."

[33]E. Cassirer, *The Myth of the State,* ECW 25, p. 40: "What matters here are not the empirical relations between causes and effects, but the intensity and depth with which human relations are felt."

[34]E. Cassirer, *The Philosophy of Symbolic Forms*, vol. 2, p. 196 (ECW 12, p. 231): "Once again we see confirmed the fundamental rule which governs all spiritual development, namely that the spirit arrives at its true and complete inwardness only by expressing itself."

[35]E. Cassirer, *The Myth of the State,* ECW 25, p. 48: "What we find here is no mere exteriorization but condensation. In language, myth, art, religion our emotions are not simply turned into mere acts; they are turned into "works". These works do not fade away. They are persistent and durable. A physical reaction can only give us a quick and temporary relief; a symbolic expression may become a *momentum aere perennius*."

Likewise, myth manifests the same tendency. This, for example, is the case of with directions—front, back, up, down, left, right—enabling orientation within sensible space[36] based on two fundamental regions, the sacred and the profane.[37] The way in which this religious distinction makes possible the differentiated occupation of that which is called to become "space" in its geometric neutrality operated by means of places first conceived as being qualitatively different. It is not further described by Cassirer, and we have yet to understand how, based on it, more regional differences such as the near and the far or the closed and the open can be introduced all the while preserving the mythical tonality from which it derives. But already, we can understand how a purely ideal norm such as the difference between the sacred and the profane can be embodied through concrete oppositions which then become carriers of *signs* and which deploy in this manner the originative mythical signification of space.

It then becomes possible to organize the mythical world based on guiding schemes, and Cassirer describes five of them: space, time, person, number, and cause.[38] As such, these guiding schemes seem well capable of being identified to *categories* inasmuch as they can be found in other forms than the mythical form, for example in scientific knowledge, in particular in what concerns causality. And this is also how

---

[36] E. Cassirer, *The Philosophy of Symbolic Forms*, vol. 2, pp. 84–85 (ECW 12, p. 99–100): "Hence in sensory as in mythical space, no "here" and "there" is a mere here and there, a mere term in a universal relation which can recur identically with the most diverse contents; every point, every element possesses, rather, a kind of tonality of its own. Each element has a special distinguishing character which cannot be described in general concepts but which immediately experienced as such. And this characteristic difference adheres to the diverse directions in space as it does to the diverse positions. [...]. In contrast to the homogeneity which prevails in the conceptual space of geometry every position and direction in mythical space is endowed as it were with a particular *accent* – and this accent always goes back to the fundamental mythical accent, the division between the sacred and the profane [ ]. The primary spatial difference, which in the more complex mythical configurations is merely repeated over and over and increasingly sublimated, is this difference between two *provinces* of being: a common, generally accessible province and another, sacred, precinct which seems to be raised out of its surroundings, hedged around and guarded against them."

[37] E. Cassirer, *The Philosophy of Symbolic Forms*, vol. 2, p. 99 (ECW 12, p. 117): "Hallowing begins when a specific zone is detached from space as a whole, when it is distinguished from other zones and one might say religiously hedged around. This concept of a religious hallowing manifested concurrently as a spatial delimitation has found its linguistic deposit in the word *templum*. For *templum* (Greek *temnos*) goes back to the root *tem*, "to cut", and thus signifies that which is cut out, delimited. It first designates then, by extension, every marked-off piece of land, every bouded field or orchard, whether it belongs to a god, a king or a hero."

[38] E. Cassirer, "The Concept of Symbolic Form in the Construction of the Human Sciences" in E. Cassirer, *The Warburg Years (1919–1933)*, p. 93 (ECW 16, p. 96): "Mythical thought has its 'categories' as does logical scientific thought. Above all, it is the fundamental and ruling category, the category of *causality*, that effectually manifests itself in it. That the most universal *concept* of causality, the simple idea of the relationship of 'cause' and 'effect' is in no way lacking in myth is clearly manifested in its content tendency to derive and 'elucidate' the world." Or also E. Cassirer, *The Philosophy of Symbolic Forms*, vol. 2, p. 44 (ECW 12, p. 54): "Mythical thinking is by no means lacking in the universal category of cause and effect, which is in a sense one of its very fundamentals."

Cassirer names them inasmuch as the mythical form indeed always deploys a *theory of the world* and that it is therefore not, in this respect, fundamentally different from the form which is scientific knowledge. But such a categorical point of view would nevertheless be misleading[39]: these guiding schemes do not have a status of category in the classical sense of the term, because it is not possible to clearly distinguish that which pertains to a category and that which pertains to another, all categories having relations of adherence and of rupture that are *relative* to one another. This is the reason why it is not possible to make of language the primitive symbolic form from which all others would be derived.

1.4.2   The Figural at the Foundation of the Indistinction Between Myth and Language

The primitive relation of indistinction of which we have just spoken is eminently present in the case of the relation between myth and language, which Cassirer examined with particular attention.[40] This examination is capital in that it shows, from the perspective of the general architecture of a philosophy of symbolic forms, that the legality specific to the symbolic form of "language" already implies that the symbolic form of "myth" is related to it in a primordial manner, each falling under the scope of the work pertaining to symbolic pregnancy.[41] Also, it is less by means of clearly circumscribed objective contents such as "language" or "myth" that it is possible to specify how symbolic forms manifest than by means of transcategorical operations which are associated to no particular content *but of which the interplay produces the typical forms of objectification.*

In the case of myth, it is the intensification of affects and their outwards projection which makes possible the constitution of entities endowed with autonomy and with their own will, such as gods or demons.[42] But this intensification and this projection would not be possible without language because it is *through the word considered*

---

[39]Cassirer indeed uses the term of mythical "category". If we make a distinction between "category" and "guiding scheme", it is to avoid misleading the reader into a theory of the univocity of predication, whereas Cassirer insists on the modalities of predication specific to each form.

[40]As we noticed already, he devoted a book to this: *Language and Myth: A Contribution to the Problem of the Names of Gods* (ECW 16, pp. 227–311) as well as a certain number of pages in the second volume of *The Philosophy of Symbolic Forms*; for example on paronymy p. 21sq; on the almighty power of the name pp. 40–42; on religious language p. 241 sq. (ECW 12, pp. 27–31; pp. 49–52; pp. 284–286).

[41]E. Cassirer, *Language and Myth: A Contribution to the Problem of the Names of Gods,* p. 205 (ECW 16, p. 303): "In the sound of language as in primarily mythical configurations, the same inner process finds its conclusion: both are the resolution of an inner tension, the presentation of psychic stirrings and arousals in certain objective formations and structures."

[42]E. Cassirer, *Language and Myth: A Contribution to the Problem of the Names of Gods,* p. 160 (ECW 16, p. 257): "[…] when external being is not simply considered and perceived but suddenly overcomes man in its sheer immediacy with the affect of fear and hope, terror or wish fulfillment, then, as it were, a spark jumps across, the tension finds a release, as the subjective excitement objectifies itself and presents itself before man as god or dæmon."

as an animated thing that such projection and intensification can take shape. There is therefore a *figural* aspect in the name employed to designate a god[43] inasmuch as the name does not designate him from the outside but *figures* within itself the activity performed by the god in question. As noted by Cassirer when referring to the research by Usener concerning the Lithuanian pantheon, gods are designated by the name of *their specific activity*: for example, the god of cattle is called the "the one who bellows" if it is the fact of bellowing which is linguistically emphasized, in the same way that the god of bees is called the "the one who buzzes" if the linguistic focus is put on the bee's buzzing.[44] Then, the mythico-linguistic form does not *have* a meaning, it *is* a meaning which becomes transposable to multiple situations,[45] thereby acquiring a certain semantic perimeter exceeding both the here and the now[46] and likely to undergo modifications in function of the directions taken by the affects and desires which always support activity.[47] *The figural is therefore, in the case of the mythico-linguistic form, the condition of possibility of its mode of objectification.*

It would now be appropriate to see in what measure it would be possible to characterize the very nature of expression from the standpoint of language, when seeking to characterize it otherwise than in the primordial kinship that language has with myth.

1.5 Idiomatic Expression in Language

We have seen earlier[48] that the place of language in the architecture of the *Philosophy of Symbolic Forms* was itself problematic and that it was necessary, in order to account

---

[43]Cassirer talks about myth as an intuitive manner of figuration. Cf. E. Cassirer, *Language and Myth: A Contribution to the Problem of the Names of of Gods*, p. 160 (ECW 16, p. 258).

[44]E. Cassirer, *Language and Myth: A Contribution to the Problem of the Names of of Gods*, p. 212 (ECW 16, p. 309): "In the table of names of Lithuanian gods given by Usener, the snow-go, the "glimmerer" *Blizgulis*, is located alongside the god of the cattle, the "roarer" *Baubis*, but also represented here are the bee-god *Birbullis*, the "buzzer", and the god of the earthquake, the 'thresher" *Drebkulys*."

[45]E. Cassirer, *Language and Myth: A Contribution to the Problem of the Names of of Gods*, p. 212 (ECW 16, p. 310): "And once a "roarer god" was conceived, he had to be recognized as one and the same being in the most diverse phenomena, he had to be heard in the voice of the lion as well as in the bellow of the storm or the thunder of the sea. Again and again, myth in this sense is revitalized and enriched by language, as is language by myth."

[46]E. Cassirer, *Language and Myth: A Contribution to the Problem of the Names of of Gods*, p. 160 (ECW 16, p. 258): "It is as if the isolation of an impression, through its being lifted out from the whole of common, everyday experience, renders with it at the same time its great intensive increase alongside a high degree of *thickening*, and, as if by virtue of this thickening, the objective figure of the god now comes about, as if it virtually sprang forth from it."

[47]E. Cassirer, *Language and Myth: A Contribution to the Problem of the Names of of Gods*, p. 163 (ECW 16, pp. 260–261): "The mode of this concentration, however, always depends here upon the direction of interest; it essentially depends not upon the content of intuition but upon de teleological perspective under which it is configured."

[48]Chap § 31.

for the integrality of Cassirer's ulterior thought, to not see in it the unique form with which all others could be put into relation but one of two forms of activity, with technology,[49] which diffracts into differentiated types of knowledge.

The role of language, in the ambiguity of its foundational role, remains, however, essential. Contrary to what occurs in the exact sciences or in the natural sciences where the explanatory level uses a system of symbols which has obviously nothing to do with what is described (planetary motions and a system of equations, for example), such is not the case with the study of language. The difficulty specific to the study of language is particularly manifest in the character, at first glance tautological, of the descriptions aiming to account for linguistic phenomena: thus, by studying the major grammatical functions, for example the notions of agency, of process, or of aspect, one may at first have the impression that the level at which the phenomena are described only manages to repeat that which is sought to be described while using other words, without a true increase in intelligibility intervening to clarify what was already known in practice. Grammatical knowledge would therefore only seek to make explicit a grammatical aptitude which is implicit at the onset for speakers, by directing their attention towards examples considered to be significant, in order to lead them to grasp what they would, in the end, be surprised to not ignore. This epistemological turn specific to the study of linguistic phenomena stems from the very unique capacity of language to become its own object, by involving within itself the intervention of various levels of intelligibility that it potentially comprises. But this is a double-edged capacity because it forces to recognize that the self-reflective power of language which so radically distinguishes itself from all other symbolic practices (scientific or artistic, for example) is also a major epistemological obstacle to its own intelligibility, because of the illusory effect of implicit transparency which is projected onto linguistic signs. Overcoming this obstacle involves to first abandon the idea that language could be studied only as an already fully achieved[50] object *when it is a mode of exploration of a potential world of objects, a world of which it also forms part itself.* The most immediate epistemological consequence of such a state of things is the unpredictable character of the forms which this exploration can take because they are not dictated by the nature of the object studied since they form, on the contrary, its possibility. This is the reason why Cassirer often expounds Humboldt's description of language as being not an *"ergon"* but an *"energeïa"*.[51] The forms of language thereby imply always novel explorations, even though, by always being

---

[49]Cassirer, "Form and Technology" p. 24 (ECW 17, p. 150): "In this respect, thinking and doing are originally united, they both stem from this common root of forming gestalts, gradually unfolding and branching off from it."

[50]E. Cassirer, *The Philosophy of Symbolic Forms*, vol. 3, p. 50 (ECW 13, pp. 55–56): "[...] The leading thought of Wilhelm von Humboldt's analysis of language is that the spiritual meaning of speech can never be fully appreciated if we consider solely the objective factor in it – if we take it as a system of signs serving solely to represent objects and their relations. [...]. Language is not taken solely as an abstract form of thought, but must be understood as a concrete form of life; it must be explained not so much by objects as by the diversity of the interpreting mood."

[51]E. Cassirer, "Structuralism in Modern Linguistics", ECW 24, p. 310: "Language is neither a mechanism nor an organism, neither a dead nor a living thing. It is no thing at all, if by this term we understand a physical object. It is – language, a very specific human activity, not describable in

available to speakers, they also account for a *specific mode of antecedence* which must be distinguished from the *a priori* such as it manifests in the exact sciences and in the natural sciences. This specific mode of antecedence must be conceived as a reservoir of meaning which, without being immediately available to the consciousness of speakers, can be reactualized depending on the circumstances.[52] The epistemological difficulty then consists in imagining the way in which to account for the antecedence that the typical forms of language—what Cassirer calls "themes"[53]—make possible, without however mobilizing a system of *a priori* categories which would also be captive to the problem of reflection—this time the reflection of the "mind"—and which would thereby let the movement specific to the constitution of still potential objects escape. By taking account of the findings of the research of his time, those by Humboldt and Boas in particular, Cassirer thus attempts to show that the idea of *a priori* categories is finally overcome in the idea of form, conceived as both a potential and actual exploration, or more exactly of which the acts, in this case language acts, manifest in the aftermath a form-generating power, an *"energeïa"*, which had preceded them.

---

terms of physics, chemistry, or biology. The best and most laconic expression of this fact was given by Wilhelm von Humboldt, when he declared that language is not an εργον but an ενεργεια. To put it shortly, we may say that language is «organic», but that it is not an «organism». It is organic in the sense that it does not consist of detached, isolated, segregated facts. It forms a coherent whole in which all parts are interdependent upon each other."

[52] An example of this specific mode of antecedence which makes linguistic expressions into an ever-exploitable repository of significations is given by Cassirer when he describes the way in which the Nazis found within the German language unexpected resources which they put into the service of their ideology. E. Cassirer, *The Myth of the State*, ECW 25, p. 279: "Not long ago there was published a very interesting little book *Nazi-Deutsch. A Glossary of Contemporary German Usage*. [...]. In this book all those new terms which were produced by the Nazi regime were carefully listed, and it is a tremendous list. There seem to be only a few words which have survived the general destruction. The authors made an attempt to translate the new terms in English, but in this regard they were, to my mind, unsuccessful. They were able to give only circumlocutions of the German words and phrases instead of real translations. For unfortunately, or perhaps fortunately, it was impossible to render these words adequately in English. What characterizes them is not so much their content and their objective meaning as the emotional atmosphere which surrounds and envelops them. This atmosphere must be felt; it cannot be translated from one climate of opinion to an entirely different one. To illustrate this point I content myself with one striking example chosen at random. I understand from the *Glossary* that in recent German usage there was a sharp difference between the two terms *Siegfriede* and *Siegerfriede*. Even for a German ear it will not be easy to grasp this difference. The two words sound exactly alike, and seem to denote the same thing. *Sieg* means victory, *Friede* means peace; how can the combination of the two words produce entirely different meanings? Nevertheless we are told that, in modern German usage, there is all the difference in the world between the two terms. For a Siegfriede is a peace through German victory; whereas a Siegerfriede means the very opposite; it is used to denote a peace which would be dictated by the allied conquerors. It is the same with other terms. The men who coined these terms were masters of their art of political propaganda."

[53] E. Cassirer, *The Philosophy of Symbolic Forms*, vol. 1, p. 269 (ECW 11, pp. 237–238): "If language is no longer regarded as a distinct reproduction of a distinct given reality, but as a vehicle of that great process in which the I «comes to grips» with the world, in which the limits of the two are clearly defined, it is evident that the problem is susceptible of many diverse solutions. For the medium of communication between I and world is not finished and determinate from the outset

We know that for Cassirer, all semiotic processes, by constituting themselves as forms, deploy their specific mode of being while generating a characteristic mode articulating signification and sense. This characteristic mode is the product of two forces acting in opposite directions: *internal coherence* which ensures the endurance of the form (its signification) and *plasticity* which ensures its adaptation (its sense). In the case of the form specific to language, internal coherence concerns the idiosyncratic character of languages whereas plasticity stems from the fact that language encompasses multiple planes of expression, for instance poetic, technical, rhetorical, and logical planes, the list of which is never closed. The difficulty with which Cassirer found himself confronted consists in maintaining a certain stability between these two poles: Cassirer must thereby show in what the "idiomatic expression"—in the very general sense of that which specifically belongs to a particular language regardless of the grammatical level with respect to which it is considered—constitutes the backbone of the internal coherence of the language all the while showing how this idiomatic expression nevertheless leaves enough leeway to allow for a receptivity to other modes of sense making—in particular the mode of logical determination which, for Cassirer, is diametrically opposed with idiomatic expression. There is therefore a double task in the description of language, indispensable in order to account for its form from the perspective of its internal dynamic, but difficult to achieve due to opposing forces which are precisely at the foundation of this dynamic.

The most immediate experience of the idiomatic expression specific to a language does not limit itself to its immutable expressions. It appears for example at the moment when we run into the difficulty of its translation: the expressive means, what we call "turns of phrases", as we know, are not the same from one language to another and it is then necessary, in order to account for signification, to mentally immerge oneself into another form—that of the language towards which we are translating—in order to attempt to manifest the same signification based on a different semantic coherence. But it is of course within a particular language that idiomatic expression manifests itself because there is no register of language that is not immediately the deployment of an idiosyncratic system of elements ensuring for each language its particular character.

In order to reveal the features specific to the idiomatic expression of language, it is also necessary to begin by discarding an almost reflex mental attitude consisting in seeing in language a logical form which has not yet fully reached its final and definite shape. As Cassirer often points out, what appears to be a defect of languages linked to

---

but comes into being and gains efficacy only by giving form to itself. Hence we cannot speak of a system, a temporal or logical progression of linguistic categories, which all linguistic development must always follow. As in epistemological inquiry, each particular category which we single out and place in relief against the others, can only be interpreted and judged as a single *factor* which may develop very different concrete configurations according to the relations into which it enters with other factors. See also E. Cassirer, *The Philosophy of Symbolic Forms*, vol. 3, p. 87 (ECW 13, p. 98): "Here again we find confirmation of our fundamental view that what we call reality can never be determined from the standpoint of material alone but that into every mode of positing reality there enters a definite motif of symbolic formation which must be recognized as such and distinguished from other motifs."

the particularity of the expressive means they deploy, or even a structural incapacity of language in general to capture the objectivity of significations,[54] is considered as such only because it is measured against a conceptual notion of signification having already acquired the univocity and universality which appears however only in logical rationality. Now, by seeing in the idiomatic aspect only an obstacle to overcome, we are condemned to miss the mode of signification inherent to the specific form of language which deploys the possibility for signification according to the mode which is specific to it and from which the logical aspect is only a derivative. In order to manage recognizing this specificity, it is necessary to leave the logical, posterior sense and to return to a more primitive sense, which Cassirer calls "teleological",[55] a sense which does not aim to describe one signification referring to properties of already constituted objects, but to describe the semiotic means employed to manage to *signify objects as objects*. The idiomatic expression is required here because no rule making possible the constitution of objects is available yet and the "teleology" of which it is question is indeed that of the finality specific to all forms as described in Kant's *Critique of Judgment*: being both its own cause and effect, it self-maintains itself by organizing the conditions of its own survival thanks to specific relations to the environment determining that which is significant for it and that which can contribute to the endurance of its own being.

---

[54]E. Cassirer, *The Logic of the Humanities*, "The Subject Matter of the Humanities", Yale University Press, 1960, pp. 80–81 (ECW 24, p. 386): "For if we are convinced that the logical concept is the necessary and sufficient condition for cognition of the essence of things, all else that is specifically different and that does not meed this standard of clarity and distinctness is only unreal appearance. In this case the illusion character of those intellectual forms which remain outside the sphere of the purely logical is indisputable. It can only be exhibited as such, and explained and justified, to the extent that we investigate the psychological origin of the illusion and attempt to set forth its empirical conditions in the structure of human representation and fantasy. The problem takes on a completely different shape if, instead of treating objects as firmly fixed in the beginning, we view them, so to speak, from an infinitely distant point, toward which all knowing and understanding aim. In this case, the "given" of objects becomes the "problem" of objectivity. And clearly it is not only theoretical knowledge which participates in this *task*; on the contrary, every energy of the mind shares in it. Language and art also are able to exhibit their particular "objective" meanings—not because they imitate a self-subsisting reality, but because they anticipate it, because they constitute distinct directions and modes of objectification."

[55]E. Cassirer, *The Philosophy of Symbolic Forms*, vol. 1, pp. 287–288 (ECW 11, p. 260): "The content of these concepts and the principle which determines their structure become fully intelligible only if beside their abstract *logical* meaning we consider their *teleological* meaning. The words of language are not reflections of stable concretions in nature and in the perceptual world, they rather indicate directions which the process of determination may follow."

1.5.1   The Expressivity of the Sentence

The most elementary idiomatic form of activity in language is the *sentence*[56]: no language limits itself to a stabilized lexicon to which syntactic rules of manipulation would be added, the distinction between lexicon and syntax coming from a division which is quite posterior to the organization of the form of language. This too artificial division fails to account for the relations of imbrication and of reciprocal influence between lexicon and syntax which languages each construct in their own manner.[57] It is, on the contrary, only on the basis of the organization of the sentence—be it reducible to a single word—that, by reflection, analysis can proceed to distinguish the lexicon and the syntax, always relatively to a particular language. Any sentence emitted by a speaker, regardless of the language, is situated therefore at a level of globality differing from that of the simple manipulation of elements which indeed compose the sentence but to which it is not reducible. Conversely, it is the necessity for organizing elements in view of ensuring the deployment of the sentence as an actualization of a figure/ground relation which implies distinguishing within it semantic roles which are differentiated roles and to thereby attribute different values to the elements which compose it. It is therefore in the reciprocal relation between the level of globality specific to the sentence and that of the means deployed to produce it that the forms of idiomaticity inherent to each language are played out. More specifically, the globality of the sentence adds nothing to the elements which compose it but only situates these at differing levels of depth by highlighting such or such element rather than another—the idiomatic aspect of the "turns of phrase" being itself conceivable at all levels of language, be it phonological, rhythmic, semantic, or syntactical. The

---

[56]What Cassirer calls the "sentence" would today be called the "utterance" in linguistics, the sentence being considered as an abstraction with respect to what is indeed being pronounced. Cf. E. Cassirer, *The Philosophy of Symbolic Forms*, vol. 1, p. 308 (ECW 11, p. 286): "Of course, in exploring this progress we must not stop at the level of mere *word formation*. The law and basic tendency of the advance are rather to be apprehended in the relations of sentence formation, for if the sentence is the true vehicle of linguistic "meaning", it is only through an investigation of the sentence that the logical shadings of this meaning can be made clear." By its very form every sentence, even the so-called simple sentence, embodies the possibility of such an inner articulation." Cassirer distinguishes himself here from the point of view of K. Bühler who endorsed the idea of the existence of a fundamental duality between the deictic and symbolic modes in the sentence. Cassirer does not however oppose this distinction since he also has recourse to it, but he does so without conceiving them as two disjoined modalities of the activity of language.

[57]Cassirer notes in particular the case of the Chinese spoken by Mongolians and by Tibetans in which the components of the sentence are often juxtaposed without apparent relations of subordination; cf. E. Cassirer, *The Philosophy of Symbolic Forms*, vol. 1, pp. 310–311 (ECW 11, p. 291): "The separate ideas that constitute discourse here lies as it were on a single linguistic plane: there is still no differentiation of perspective between foreground and background in speech itself. Language reveals its power of differentiation and articulation in the coordination of the parts of the sentence; but it does not yet succeed in raising this purely static relation to a dynamic relation of reciprocal logical dependency, and expressing it as such. In place of precisely graduated subordinate clauses, a simple gerundial construction may serve, without departing from the law of coordination, to express the most diverse specifications and modifications of action, encompassing them in a stable, but characteristically rigid construction."

highlighting of elements inevitably relegates others to the background and it is these differentiated phases which enable to manifest the phenomenon of signification.[58] Therefore, the sentence manifests what has the vocation of becoming an object by distinguishing, in its own process, phases in which the figure/ground relations are idiosyncratically articulated.

Directed towards another person, the sentence as it is deployed in discourse already appears as the bearer of idiomatic expressivity at the phonological level. We immediately experience this in the diversity of languages or in the regional and social accents of a same language. This idiomatic expressivity, which we could believe to be entrenched at this level and which we could thereby attribute to a nature specific to sounds themselves,[59] is nevertheless immediately the bearer of other determinations at other levels, which we will distinguish for practical reasons as pertaining to semantics (expressing for example proximity, slowness, or swiftness) or of syntax (for example the difference between a proposition under its affirmative or negative forms)[60] without it being possible to predict which feature will be drawn from the expression in a given language and at a given grammatical level. The natural character specific to expressivity—which Cassirer does not seek to deny—therefore also immediately ceases to be the foremost characteristic by integrating itself into the idiomatic expression of a particular language, that is, from the moment it acquires a linguistic value as such.[61] This process draws its strength in a manner internal to language without there being a necessity for supposing the existence of a "nature"

---

[58] Cassirer generally distinguishes three phases which he describes as "mimetic", "analogical", and "symbolic". Cf. E. Cassirer, *The Philosophy of Symbolic Forms*, vol. 1, p. 190 (ECW 11, p. 137): "In general, language can be shown to have passed through three stages in maturing to its specific form, in achieving its inner freedom. In calling these the mimetic, the analogical, and the truly symbolic stage, we are for the present merely setting up an abstract schema – but this schema will take on concrete content when we see that it represents a functional law of linguistic growth, which has its specific and a characteristic counterpart in other fields such as art and cognition."

[59] E. Cassirer, *The Philosophy of Symbolic Forms*, vol. 1, p. 193 (ECW 11, p. 141): "In the same way certain consonants and consonantal groups are used as "natural phonetic metaphors" to which a similar or identical significatory function attaches in nearly all language groups – e.g., with striking regularity the resonant labials indicate direction toward the speaker and the explosive lingual direction away from the speaker, so that the former appear as a "natural" expression of the "I", the latter of the "Thou"."

[60] E. Cassirer, *The Philosophy of Symbolic Forms*, vol. 1, p. 194 (ECW 11, p. 141): "These variations serve as a basis both for etymological distinctions – i.e., the same syllable serves, according to its tone, to designate entirely different things or actions – and for spatial or quantitative distinctions, i.e., high-pitched words, for example, express proximity and slowness, etc.... And purely formal relations and oppositions can be expressed in this same way. A mere change in tone can transform the affirmative into the negative form of a verb. Or it may determine the grammatical category of a word; for example, otherwise identical syllables may be identified as nouns or verbs by the manner in which they are pronounced."

[61] E. Cassirer, *The Philosophy of Symbolic Forms*, vol. 1, p. 180 (ECW 11, p. 125): "Every elementary expressive movement does actually form a first step in spiritual development, in so far as it is still entirely situated in the immediacy of sensuous life and yet at the same time go beyond it. It implies that the sensory drive, instead of proceeding directly towards the object, encounters a kind of inhibition and reversal, in which a new *consciousness* of this same drive is born. In this sense the reaction contained in the expressive movement prepares the way for a higher stage of action."

linked to an already constituted "object": Cassirer uses, as an example, the formation in German of the suffix -*keit* which appears to form a whole to which we could believe that a nominalizing function has always been attached since it constitutes one of the clearly identified sets of substantives of the feminine gender in this language. In fact, this is not actually the case since, as Grimm has shown, the construction of the suffix -*keit* results from the amalgam between two phonological traits having between themselves for sole relation that of proximity in the spoken sequence— and thereby having no vocation to express anything whatsoever—but of which the fusion is nevertheless immediately invested with a particular grammatical value.[62] This phonological amalgam is exemplary of the intervention of the form specific to language upon the sound material: capable of bringing the most disparate elements together to make them into carriers of value at all levels of grammatical analysis, language nevertheless constructs in this manner a viable order, that of the sentence, founding the relation of signification.

It remains that the way in which language manages, from disparate elements, to make perceptible in speech differentiated semantic roles at grammatical levels which are themselves differentiated remains, as such, enigmatic, because we dot not yet grasp the profound reason even if we perceive its character which seems irrational at first glance—as the example of the suffix—*keit* has just demonstrated.

1.5.2   Perception Under the Aegis of Language

Laterally, if we could say so, "objects" are therefore specified bearing the trace of the idiomatic activity which presided over their linguistic construction. Thus, for example, Cassirer uses on several occasions an example by Humboldt[63] describing

---

[62]E. Cassirer, *The Philosophy of Symbolic Forms*, vol. 1, pp. 307–308 (ECW 11, p. 286): "At first sight, to be sure, suffix formation might seem to be essentially characterized by the fact that the original substantial signification of the word from which the suffixes derive, is progressively thrust into the background and ultimately forgotten altogether. This forgetfulness sometimes goes so far that new suffixes may arise, which owe their origin to no concrete intuition but to what one might call a misguided impulsion of linguistic analogy formation. In German, for example, the suffix -*keit*, goes back to a linguistic "misunderstanding" of this sort: in words such as *ewic-heit*, the final *c* of the stem blended with the initial *h* of the suffix, so as to form a new suffix which was propagated by analogy. From a purely formal and grammatical point of view, such processes are regarded as "aberrations" of the linguistic sense; actually they are far more, they represent *progress* to a new formal view, a development from substantial expression to the expression of pure relation."

[63]E. Cassirer, *The Philosophy of Symbolic Forms*, vol. 1, pp. 284–285 (ECW 11, pp. 256–257): "For the word is not a copy of the object as such, but reflects the soul's image of the object. In this sense, the words of different languages can never be synonyms – their meaning, strictly speaking, can never be emcompassed in a simple definition which merely lists the objective characteristics of the object designated. There is always a specific mode of *signification* which expresses itself in the syntheses and coordinations underlying the formation of linguistic concepts. If the moon in Greek is called «measurer» *men*), in Latin the «glittering» (*luna, lucna*), we have here one and the same sensory intuition assigned to very different notions of meaning and made determinate by them. It no longer seems possible to give a general account of the way in which this specifying of intuition is effected in different languages, precisely because we have to do with a highly complex cultural

the way in which, in different languages, a same entity—the moon—is constituted into an "object" on the basis of construction activities which have nothing in common: the moon is designated as "that which measures" in Greek (*mên*) whereas in Latin, it is its "brightness" which is emphasized (*lucna*). The expressivity specific to the moon and the capture of one of its expressive quality enables to converge towards an object serving as referent, the moon. That such or such language is sensible to the measurement which can be induced from lunar cycles, and that such or such other language is sensible to its brightness seems completely arbitrary. Changing constructions to designate the same object would obviously only stave off the problem because, each time, there would be no way to name an object without idiomatically distinguishing a particular characteristic which would be attached to it by one means or another, even if remote. The example taken from Humboldt is therefore capital for two other reasons.

One the one hand, the idiomatic constructions, whatever they may be, are not posterior to the grasping of the object, but it is what makes the perception of the object possible since, without them, the entity would not be susceptible of being designated as an object. Perception *institutes*, thereby, an order going beyond that which pertains only to sensation, a simple chaos of impressions still too undifferentiated to distinguish objects.[64] From this must therefore be concluded that the *idiomatic constructions of language found the very possibility of perception* which is not limited to a silent grasp of already clearly distinguished entities but which, on the contrary, collects and organizes an interlacing made of geometric contours and of qualitative intensities grasped by a corporeal or technical apprehension in a linguistic idiomaticity, in view of producing determined objects. It is this interlacing that Cassirer names "symbolic pregnancy"[65] and which makes the nature of perception so difficult to

---

process, which varies with each case."; cf. also E. Cassirer, *An Essay on Man, An Introduction to a Philosophy of Human Culture*, ECW 23, p. 145: "As Humboldt pointed out, the Greek and Latin terms for the moon, although they refer to the same object, do not express the same intention or concept. The Greek term (men) denotes the function of the moon to "measure" time; the Latin term (Luna, luc-na) denotes the moon's lucidity or brightness. Thus we have obviously isolated and focused attention on two very different features of the object."

[64] E. Cassirer, *The Philosophy of Symbolic Forms*, vol. 1, p. 87 (ECW 11, p. 18): "The process of language formation shows for example how the chaos of immediate impressions takes on order and clarity for us only when we "name" it and so permeate it with the function of linguistic thought and expression."

[65] E. Cassirer, *The Philosophy of Symbolic Forms*, vol. 3, p. 202 (ECW 13, pp. 230–231): "It is with a view to expressing this mutual determination that we introduce the concept and the term "symbolic pregnance". By symbolic pregnance we mean the way in which a perception as a sensory experience contains at the same time a certain nonintuitive meaning which it immediately and concretely represents. Here we are not dealing with bare perceptive data, on which some sort of apperceptive acts are later grafted, through which they are interpreted, judged, transformed. Rather, it is the perception itself which by virtue of its own immanent organization, takes on a kind of spiritual articulation – which, being ordered in itself, also belongs to a determinate order of meaning. In its full actuality, its living totality, it is at the same time a life "in" meaning. It is not only subsequently received into this sphere but is, one might say, born into it. It is this ideal interwovenness, this relatedness of the single perceptive phenomenon, given here and now, to a characteristic total meaning that the term "pregance" is meant to designate."

grasp because we tend to suppose that it can always be divided into more primitive components, thus contributing to the belief in the reality of an antepredicative and "natural" state of that which is perceived. Now, the interlacing of that of which it has just been question shows that, in perception, there are in fact several levels of intelligibility which can indeed be subsequently distinguished, but that perception is not at the onset a uniform construction even if it is possible, by means of lengthy work, to isolate in it certain universal constraints which are not linguistic as such and which are essentially of a geometrical and dynamical nature.[66] These constraints, once isolated, do not however form the ultimate substrate of perception because they have the serious shortcoming of being *mute* and they will thereby tend to occult the primary role of language in the elaboration of perception.

On the other hand, the example given by Humboldt shows that the characteristics of the idiomatic constructions also proceed as much from sensible data (brightness in Latin) as from mereological data of a cultural nature (measurement in Greek) without any chronological or logical anteriority being establishable between the two types of constructions.[67] As we have seen already, language resists any unilateral confinement and there is no way it could be ordered in advance in categories.[68] Indeed, if in this particular case, Latin appears closer to immediate qualitative sensibility whereas the Greek version seems to require a more mediate construction, this does not however say anything about the other idiomatic construction which can be found in both languages nor about the possibility of a hierarchical classification of languages in general from the "most sensible" to the "least sensible" because all types of possible constructions are received in a language. It should be noted here that Cassirer does not defend the idea of an absolute value scale which would enable to hierarchically classify languages according to their construction types and that even if he recognizes, following Humboldt, the particular force of the notion of inflection, he does not conclude from it, as did Humboldt, that the inflectional languages would represent some acme of linguistic activity[69] because it is always with respect to a

---

[66]Cassirer developed a strong interest for the Berlin school of the psychological theory of Gestalt and he abundantly cites Köhler and Wertheimer; cf. for example E. Cassirer, *The Philosophy of Symbolic Forms*, vol. 1, respectively p. 189 and p. 233 (ECW 11, p. 136 and 210).

[67]E. Cassirer, *Substance and Function*, p. 222 (ECW 6, p. 242): "For instance, there is no entirely definite intuitive content corresponding to the word "bird", but rather only a certain vague outline of form along with a vague presentation of a wing movement, so that a child may call a flying beetle or butterfly a bird; the same is originally true of all our universal presentations. There are only possible because we have, along with the concrete and complete sense perceptions, also perfect and definite contents of consciousness. The indefiniteness of the memory-images of our actual sensations involves that, along with the vivid and immediately present, sensuous intuitions in the real process of consciousness, pale residua of them are always found, which retain only one feature of them; and it is these latter, which contain the real psychological material for the construction of the universal presentation."

[68]Cf. supra 'The Symbolic Turn': 'The ambivalence of the place of language in the architecture of Symbolic Forms'.

[69]E. Cassirer, *The Philosophy of Symbolic Forms*, vol. 1, p. 309 (ECW 11, pp. 287–288): "The differentiation of the word and its integration with the sentence form correlative methods which join in a strictly unitary operation. Humboldt and the older philosophers of language looked upon

given interpretive framework that a notion is considered as being more fundamental than another.

This point allows to clarify the relationship which Cassirer had with the notion of "primitivism" in linguistics, which must be radically distinguished from idiomatic expression.

1.5.3 Primitivism

It is difficult not to be startled by the appearance, under Cassirer's pen, of expressions pertaining to an anthropology which today would seem outdated such as the "language of primitive peoples",[70] or even the abridged expression "'primitive' languages",[71] as opposed to "highly developed languages",[72] expressions which appear to consecrate a hierarchy between peoples and their languages, without it being clear what is the nature of the link uniting the languages and the "backward" or "advanced" state of the societies to which they are associated.

To understand Cassirer's point of view, it is first necessary to start with a brief remark on his vocabulary. Cassirer mostly uses two substantives, in plural form, but the first one always with quotation marks: "*Primitiven*" and "*Naturvölker*". In the German context of the 20s, the term "*Primitiv*" could be a sign of the influence of Anglo-saxon and French anthropologies in which the expressions "primitive peoples"[73] or "primitive mentality"[74] were commonly used. As for the second one, the term "Naturvolk" is more specific to the German context and could be translated by "indigenous people" if the conventions of the time were to be followed. This last expression has both positive and negative connotations: positive in the sense that it

---

this circumstance as a proof that the true inflected languages represent the summit of all language formation and that in them, and only in them, the "absolutely lawful form" of language expressed itself in ideal perception. But even if we show a certain skepticism and reserve toward such absolute evaluations, there is no doubt that the inflected languages provide a highly important and effective organ for the development of purely *relational thought*."

[70] E. Cassirer, *The Philosophy of Symbolic Forms*, vol. 1, p. 199, p. 217 (ECW 11, p. 137, p. 170). Here, Cassirer does not use the adjective 'primitive' as it is rendered in the English translation. The expression he uses is 'the languages of natural peoples' [*die Sprachen der Naturvölker*]. When Cassirer makes use of the term 'primitive' [*Primitiv*], he uses quotation marks; see next quote.

[71] E. Cassirer, *The Philosophy of Symbolic Forms*, vol. 1, p. 223 (ECW 11, p. 178). Cassirer speaks of "the grammar of 'primitive' people" [*Grammatik «primitiver» Sprachen*] and of the 'languages of the 'Primitives' [*Sprachen der «Primitiven»*].

[72] E. Cassirer, *The Philosophy of Symbolic Forms*, vol. 1, p. 190 where Cassirer uses the expression 'developed languages' [*entwickelten Sprachen*] or p. 314 'highly developed languages' [*hochentwickelten Sprachen*] (ECW 11, p. 137, p. 294).

[73] See for example D. G. Brinton, *Religions of Primitive Peoples*, New York 1897, whom Cassirer quotes an article in *The Philosophy of Symbolic Forms*, vol. 2, p. 275 note 210 (ECW 12, p. 245 note 220).

[74] See for example L. Lévy-Bruhl, *Les fonctions mentales des sociétés inférieures*, Paris, 1910 quoted by Cassirer in its German translation *Das Denken der Naturvölker* in *The Philosophy of Symbolic Forms*, vol. 2, p. 230 note 85 (ECW 12, p. 188 note 87).

refers to the notion of origin which was highly praised at that time and negative in the sense that it introduces a hierarchy between peoples, if only indirectly.

That said, one must notice that nowhere in Cassirer the principles governing such a classification between peoples or languages can be found, if not negatively, when Cassirer on the contrary denounces the distinction between inferior and superior such as it may be found in some racist tendencies of the linguistics of his age[75] (hence his use of the term "primitive" with quotation-marks). The racist prejudices of the time in both anthropology and linguistics are numerous, but those against which Cassirer fought in particular in the field of linguistics concern the idea that so-called "inferior" languages would be no more than simple tools enabling the empirical accumulation of facts without bearing any conceptual objective. In fact, *there is no language which is not replete with conceptual potentiality*. Cassirer is, in this respect, very clear, as the example concerning counting in Sotho shows: it is indeed a Bantu language which he solicits—and not, for example, Greek, despite its being so strongly linked with the origin of mathematics in Western culture—in order to address such a highly conceptual issue related to the conception of counting and therefore does not relegate it to an inferior status. The counting procedure in Sotho is indeed body-dependent since it is made possible by counting one's fingers then one's toes but, regarding another theme than counting, that of the construction of the word "moon", the same thing could be said about Latin—a supposedly "evolved" language—which constructs *"lucna"* based on the most immediate qualitative sensibility of brightness. In the case of Sotho, Cassirer remarks on the contrary that the specifically conceptual dimension of the act of counting bases itself on the more empirical level of one's own body only to better exceed it because it is the *necessity* for sequencing which constitutes the act's foundation regardless of the physical medium but always in conjunction with it. As for the second case, the "immediateness" of the Latin construction for "moon" is not to be opposed to the "mediate" Greek construction because it is easy to imagine that, regarding another word, it is the Greek version which appears more "immediate"—in the sense of being closer to the sensibility of the one's own body—than the Latin expression.

The notion of "primitive" therefore designates, for Cassirer, a state of languages which is not primitive in the naïve and racist sense of the term[76]; it designates a state of

---

[75]E. Cassirer, *The Philosophy of Symbolic Forms*, vol. 1, pp. 230–231 (ECW 11, p. 188) which cites in particular Steinthal, *Die Mandeneger Sprachen*, Berlin, 1867: "The intellectual value of primitive counting methods has often been disparaged. […]. But in the half-poetical, half-theological pathos of his diatribe, Steinthal forgets that it is far more fruitful to seek out and to recognize the intellectual content of this method however slight than to measure it by our fully developed concept of number."

[76]What distinguishes Cassirer's attitude from Steinthal's (racist) attitude is that Cassirer never considers the state of a language as *definitively* set in a "primitive" posture, even if such is the way it may present itself in the current state of its description. This dynamic character must therefore be kept in mind when reading a certain number of texts which could otherwise lead to confusion; for example E. Cassirer, *The Philosophy of Symbolic Forms*, vol. 1, p. 190 (ECW 11, p. 137): "Here the sound seeks to approach the sensory impression and reproduce its diversity as faithfully as possible. This striving plays an important part in the speech both in children and "primitive" peoples. Here language clings to the concrete phenomenon and its sensory image, attempting as it were to exhaust it in sound; it does not content itself with general designations but accompanies every particular

language which is just as relevant for describing the languages which were at the time called "evolved" but which does not manifest, in such or such particular occurrence, its *propensity for meaning*, that is, its *potential for symbolic transformation*. It is the opposition between a form's internal signification and sense external to it which enables here to clarify this point. A particular feature which is only considered from the standpoint of its signification, that is, from the point of view of the place it occupies within the system internal to a language, is considered as primitive because it is considered for itself, in the place it occupies within the linguistic structure, without a potential for symbolic transformation. The same feature will be on the contrary considered to be evolved when it is considered to open towards new forms and when its potential for symbolic transformation is recognized.

This is the case with counting in Sotho which, in Cassirer's view, manifests its propensity for sense (and not only for signification) when it introduces the necessity of the rule making well-ordering possible and that it thereby enables to construct the *concept* of well-ordering, indispensable for conceiving the mathematical definition of number, as found by Cassirer in the works of R. Dedekind in particular.[77] Likewise, the copula in Indo-European languages manifests, for Cassirer, its potential regarding symbolic transformation because it typically introduces to the notion of relation in its logical form and not only in its linguistic form; the notion of relation then appears to be exemplary—in the Goethean sense—in the propensity it has for representing, in the sphere of language, a meaning which *extends beyond this sphere* and which opens onto a new form—here, that of pure logic.[78] What is here called "potential for symbolic transformation" therefore describes a semiotic moment of transformation in which a particular feature is envisioned both in its signification and in its sense, that is, both as a reality *internal* to a form and as a potential for transformation that is *external* to this very form—in its relation to other forms. This moment, of a dynamic

---

nuance of the phenomenon with a particular phonetic nuance, devised especially for this case. In Ewe and certain related languages, for example, there arc adverbs which describe only *one* activity, *one* state or *one* attribute and which consequently can be combined only with *one* verb." The parallel between children and "primitives" could be seen as an echo to a racist cliché if it were to be taken to the letter. As contestable as it may be—and it is necessary to criticize Cassirer in this respect since he suggests that such a parallel is justified—, this parallel must nevertheless be interpreted in dynamic terms, Cassirer always insisting on the integral unity of human behaviors throughout space and time: in his view, there is no absolute primitivism.

[77] E. Cassirer, *Substance and Function*, p. 53 (ECW 6, p. 54): "The theory of the ordinal number thus represents the essential minimum, which no logical deduction of number can avoid; at the same time, the consideration of equivalent classes is of the greatest interest for the application of this concept, yet does not belong to its original content."

[78] Another example is that of "vowel harmony" which Cassirer shows to be a fundamental *phenomenon* of construction in Ural-Altaic languages but which also holds as a *principle* for language in general; cf. E. Cassirer, *The Philosophy of Symbolic Forms*, vol. 1, pp. 194–195 (ECW 11, p. 143): "We are carried one step further by the phenomenon of *vowel harmony* which dominates the whole structure of certain languages and linguistic groups, particularly those of the Ural-Altaic family. [...]. In becoming a phonetic unit through the principle of vowel harmony, the word or word-sentence gains its true significative unity: a relationship which at first applies solely to the quality and physiological production of the particular sounds, becomes a means of combining them in a spiritual role, a unit of "signification"."

nature, thus defines generative instabilities which serve as global constraints upon the general system of symbolic forms and their relationships.

It would be very natural, at this point, to show how idiomatic expression in language immediately generates, through the potential of symbolic transformation inherent to language, a fabric of relations which fully reconfigures it and conduces it to make itself into the bearer of a necessity of which it first appeared devoid: it is also in this manner that Cassirer proceeds in *The Philosophy of Symbolic Forms* since he demonstrates therein how the notion of relation progressively moves from the idiomatic towards the conceptual. However, as we have already remarked,[79] the schema with which Cassirer begins in that text does not account for the ulterior evolution of his own thinking, from the moment when he involves in his reflection other sorts of activities—such as technological activity, for example—or other forms of knowledge—such as esthetics. Also, it appeared more expedient, in order to respect this evolution, to involve the notion of "transcategorical operation" which must account not so much for the types of activities or types of knowledge than for the potential for symbolic transformation belonging to all forms—be they activities or knowledge—regardless of their nature or number. We will however retain from the presentation of *The Philosophy of Symbolic Forms* the idea that the activity of language as well as mythical and scientific knowledge are indeed the three most paradigmatic themes for describing idiomatic expression, evocation, and objectification, and it is therefore by beginning with them that we will introduce these notions.

This point leads us to now address the question of idiomatic expression in the form of scientific activity, thereby showing that the idiomatic expression which Cassirer describes in language is only one aspect, indeed fundamental, of a more general operator which is deployed through all symbolic forms and of which the development is not achieved according to a linear schema of progression which would only go from language to science passing through myth.

1.6 Idiomatic Expression in Science

We will here insist less on the case of expression in science than we will on the preceding cases of expression in myth and language because, on the one hand, we have on several occasions provided examples of this expressivity—in particular, the occult link, prior to the work of philology, which existed between Aristotelian logic and the Greek language, or the "natural" character of Euclidean geometry—and, on the other hand, we have justified the attitude of science with respect to expression in the context of a philosophy of symbolic forms when we described the propensity of science to attempt to unbind itself from any relation to expression. We would rather

---

[79]Cf. Chap. 3, § 2, where we already noted that the evolution of Cassirer's thought led him to consider the generation of symbolic forms as the result of three transcategorical operators which have been called "expression", "evocation", and "objectification".

wish to conceive of the nature of the semiotic link which nevertheless continues to exist between science and expression and to then give, regarding the particular case which Cassirer calls the "pathology of symbolic consciousness",[80] the example of such a link.

1.6.1 The Retrospective and Prospective Movement of Expression in Science

We have already remarked on several occasions that science was constituted in rupture with the expressivity of language and of myth, at a moment which Cassirer qualifies as a "crisis". From a certain point of view, however, this crisis is indefinite because the propensity to expel any expressive resource is at the very origin of the possibility for a history of science inasmuch as science targets the concept. But this does not mean that the divorce between science and expression is total because the expelling is *also* a specific modality of a semiotic relation to expression. This specific modality is characterized by the fact that the *relation to expressivity is undergone by science on the mode of the overcome*. It is therefore always in a manner that *it wishes to be only retrospective* that science considers its possible links to expression. Such a relation to expression induces an impoverished conception of it as being only an obstacle—a passive "substrate" or inert "matter"—which might haunt the past of science but which does not burden its future.

In fact, the epistemological work of Cassirer has shown that such was not the case. Not only had he properly established the existence of a retrospective link between expression and science given that the opposition between substance and function enabled him to distinguish ancient and medieval science on the one hand from modern and contemporary science on the other in all sorts of manifestations over the course of history, but he went even further: the uncontrollable plurivocity of sense continues to affect the very advancement of science as namely proven by the examples related to the philological revolution of the Renaissance. The tendency of the symbolic form which is science therefore is, as is for any symbolic form, to attempt to fully occupy the terrain of sense by seeking to devitalize its plurivocity to the sole benefit of signification's conceptual univocity. But the example of the increasing generalization of the function in all the natural sciences since the seventeenth century has nevertheless finished by otherwise making the relation to sense resurface *by making plurivocal* the geometric norm itself. Therefore, it is henceforth at the very core of the advancement of science that is situated the relation to the plurivocity of sense and, from this perspective, the relation to sense and to signification is always to be renegotiated, in the natural sciences as in the sciences of culture.

---

[80]E. Cassirer, *The Philosophy of Symbolic Forms*, vol. 3, p. 205 (ECW 13, p. 238).

1.6.2   The Normal and the Pathological

The case of the normal and the pathological provides an example. Pathological behaviors are interesting with respect to the relation between expression and science inasmuch as they reveal, using concrete clinical examples, that no science can develop without previously founding itself upon what linguistic and perceptive expression puts at its disposal. This is also the reason why the patients in question need to adopt convoluted strategies in order to successfully carry out spatial or counting tasks that normal subjects perform without thinking. From the more general standpoint of the relations between science and expression, it is therefore only once the expressive basis of perception and of language has been ensured that symbolic work on the stabilized forms becomes possible and that the construction of concepts as important as those of uniform space, of direction, or of counting becomes possible. Cassirer uses the example of the numerous pathologies which affect both the perception of space and of time and which thereby implicitly sketch out the geometrical and arithmetic norms. He cites in particular the case of a patient who had lost the capacity to trace the main directions of orientation: above, below, right, and left.[81] With respect to the temporal relations having a direct relationship with the conception of the ordinality and cardinality of numbers, the patient could tell the sequence of days and of months without being capable of stating the day or the month following or preceding the ones he was given. The same patient was able to recite the sequence of natural numbers but was unable to count objects using these very numbers, as if the meaning of their cardinality was lost on him. Cassirer relates these difficulties pertaining to space and to counting to aphasia as a linguistic pathology, because only language enables the prior fixation of elements which can later be permutated in the performance of an operation. Cassirer gives the example, very familiar in the Kantian tradition, of the addition of 7 and 5 and shows that the additive operation first consists in the determination of elements set in language which then occupy a position within the series of numbers and that later, the arithmetic operation as such consists in varying the origin from which the series is begun: therefore, adding 5 and 7, once the unit has been counted seven times, consists in shifting the origin of the count to that point, that is, to make it occupy the position of zero, and then reiterate the process five times.[82] The capacity

---

[81] E. Cassirer, *The Philosophy of Symbolic Forms*, vol. 3, p. 247 (ECW 13, p. 286).

[82] E. Cassirer, *The Philosophy of Symbolic Forms*, vol. 3, p. 250 (ECW 13, p. 289): "For every such act requires not only that the number be posited at this or that one, as a determination within a series, but at the same time that this positing of unities can be freely varied. It not only requires correlation with the numerical series as a fixed schema; this schema despite its fixity must be regarder as mobile. The nature of the union of the two seemingly contrary requirements and the way in which it is achieved, are shown by every elementary example in addition or substraction. Essentially, to find the sum 7 and 5, or the difference between them, means nothing more than to count five steps forward from 7 or five steps backward. Thus the decisive factor is that the number 7, although retaining its position in the original series, is taken in a new meaning, as the starting point of a new series where it assumes the role of zero. Every number in the original series can thus be made into the starting point of the new series. Now the beginning is no longer an absolute beginning, but a relative one: it is not given but must be posited in each case according to the conditions of the problem. Thus the difficulty presents a perfect analogy to the difficulty we have seen in the

to make mobile the point of origin for counting units considered to be linguistically set thus enables to execute an arithmetic operation based on a *geometric* permutation and this is the reason why Cassirer can establish an arithmetic and geometric relation using this very simple example: as the geometry of space is lacking in the patient who is incapable of displacing the origin of spatial coordinates, likewise is it the displacement of the numerical origin, that is, taking into account its only *relative* aspect, which lacks when it is a matter of executing a simple arithmetic operation. Therefore, language and the geometry of perception appear indeed as conditions *sine qua non* for the advent of an aim having for norm the schematization of science.

It therefore clearly appears that the relations of science and expression do not have the simplicity one would spontaneously tend to confer them and that there are among them links which are most of the time hidden and which the framework of a philosophy of symbolic forms seeks to study in detail. We must now address the question of expression in a symbolic form which Cassirer did not take into account when he wrote *The Philosophy of Symbolic Forms*, that is, technology, to which he gave attention right after the final publication of this work, in an article dating of 1930.

1.7 Idiomatic Expression in Technology

Technology appears to Cassirer as the symbolic form which is the most primordial among human activity as it is the result of the will and not of the power of desire which, for its part, is linked to the forms of myth and of magic. Technology is, from this perspective, much more than a simple solution to a practical problem: it may be viewed as a new order of the world, as a new form of its objectification. A difficulty immediately arises concerning the nature of idiomatic expression specific to technology because it first appears as *mute*, as Cassirer says citing one of his contemporaries.[83] Technical activity indeed has the inherent tendency of "going unnoticed" and of producing solutions of which the practical implementation may be entirely delegated to the technological device involved. Thus, there is no need to remember the electrical circuit which is nevertheless indispensable for turning on the light, nor to fully grasp a motor's inner workings in order to drive a car. In such conditions, what could be the form of expression specific to technology if the practical setup which presides its functioning seems designed to be devoid of expression?

Cassirer remarks that some have attempted to answer this question by considering technology to be an emanation of the body conceived as the primary dimension of any idiomatic expression.[84] But he criticizes this viewpoint inasmuch as technology

---

perception of space: it consists in the free positing and free removal of a center of coordinates, and also in the transition between systems based on different centers."

[83] He cites Max Eyth in "Form and Technology", pp. 22–23 (ECW 17, pp. 148–149).

[84] Such is the case, Cassirer tells us, of Kapp; cf. "Form and Technology", p. 37 (ECW 17, p. 167).

precisely does not limit itself to the body because its intrinsic tendency is, on the contrary, to gain independence from the body. It therefore cannot be only the *imitation* of the function of an organ by other physical means. Rather, technology is an idiosyncratic activity which imitates no model exterior to itself. An example of this may be found with flight technology, in which the form of the solution to the technical problem of flight will be truly addressed when ceasing to want to build the wings of an airplane by imitating the wings of a bird using a mobile wing. In doing so, the technical activity as such reveals its close kinship with the physical construction of the world,[85] and it then becomes possible to suppose, even if Cassirer does not say so expressly, that expression, as it manifests in science and which he described as resting first and foremost on the distinction between substance and function, could play an analogous role in the case of technology. It would be suitable then to distinguish between a substantial expressivity of technology and a functional one, in accordance with the degree of conceptual integration which could be performed on the elements participating to the technical apparatus.[86] "Substantial" technology would still base itself on an imitation of natural processes and would keep with the perimeter of action they make possible, sometimes also conferring it a magical force.[87] On the other hand, "functional" technology would involve that the interaction of elements participating to the apparatus would be based on variation and adaptability to all possible cases. Such is the case with flight which Cassirer uses as an example without going into details: the imitation of the wing of a bird in the form of a mobile wing tries to adapt a type of flight to conditions which are not generally reproducible. By breaking with such a thoroughly imitative model, the wings of the plane constitute a functional response which no longer limits itself to the perimeter of validity of a bird's wings alone but which responds to all possible cases regardless of the weight or size of what needs to be put into flight. Yet, Cassirer's example describes a technological failure, that of the mobile wing, and does not enable to grasp what would be a working substantial technical apparatus. We could imagine another example, such as the following: faced with the problem of making the displacement of loads in space as easy as possible, we could attempt to respond in a substantial manner

---

[85] E. Cassirer, "Form and Technology", p. 42 (ECW 17, p. 174): "For it is in no way the 'abstract', pure technological knowledge of the laws of nature that leads the way, proving first the technological aspect of the problem and its concrete technological activity. From the very beginning, both processes grasp one another and, as it were, keep the balance."

[86] E. Cassirer, "Form and Technology", pp. 38–39 (ECW 17, p. 169): "As to the basic principle that rules over the entire development of mechanical engineering, it has been pointed out that the general situation of machines is such that they no longer seek to imitate the work of the hand or nature, but instead seek to carry out tasks with their own authentic means, which are often completely different from natural means. Technology first attained its own ability to speak for itself by means of this principle and its ever-sharper implementation."

[87] E. Cassirer, *Language and Myth: A Contribution to the Problem of the Names of of Gods*, pp. 180–181 (ECW 16, p. 278): "However, here, too, it may be observed that, as soon as the human being employs the tool, it is no longer a mere product in which he knows and recognizes himself as its creator. He sees in it not a mere artifact but an independent being, something that possesses its own powers. Instead of ruling it with his will, he turns it into a god or a dæmon on whose will he depends and to which he feels himself subjugated and which he religiously and ritually venerates."

by training animals (for instance horses) in order to achieve this aim—the technological apparatus would then reside in training techniques—but one could as easily
respond in a functional manner using the technology of the wheel which, although it
is nowhere to be seen by the naked eye in nature, resolves in a perfectly adapted manner the displacement problem while distinguishing it from the problem of traction.
The tendency towards technical adaptation would then be the manifestation of an
ever-increasing integration of the elements present in technological devices within a
positively *theoretical* plane which radically modifies the solely sensible relationship
to nature. Technology thus pertains to a form having no preceding equivalent to it
and can therefore be legitimately classified among symbolic forms.

For Cassirer, the tool, forming a microcosm of all technology, is neither characterized by the goal it tries to accomplish nor by the utility which stems from its usage,
both being exterior to its specific form: the tool, and therefore also, more generally,
technology, manifests the form of the *activity* itself as it supposes an intrinsic work
on its form. As Cassirer remarks, action does not only involve action *using* tools
but also performing an action *upon* tools[88] and it is in this work upon the tool as
such that its expression is manifested.[89] We can then epistemologically compare this
form of expression with those with which other symbolic forms are endowed: the
relation between technology and art, between technology and science, and finally,
between technology and ritual. Whereas art seeks expression *for its own sake* under
the aspect of beauty,[90] technological expression essentially resides in the conceptual *autonomy* of devices which technology deploys independently from the subjects
which produced them, in a process which seeks to keep solely with the sphere of
objectivity.[91] But while the expression of science, which also keeps with the sphere
of objectivity, is of an exclusively theoretical nature, technological expression has
this in common with art that it is indeed within sensible matter that its theoretical
ambition manifests: its conceptual autonomy is transmitted to the material devices
it implements. From a genetic standpoint, finally, technology first appears as the

---

[88] E. Cassirer, "Form and Technology", p. 33 (ECW 17, pp. 162–163): "The basic force of the human
being reveals itself perhaps nowhere as clearly as in the sphere of the tool. The human works with
it insofar as he, in some way, even if initially with only modest results, works on it."

[89] E. Cassirer, "Form and Technology", p. 32 (ECW 17, p. 161): "The object is determined as
something in so far as it is for something. This is because in the world of tools there are no mere
things with properties. There are only ensembles of 'vector-magnitudes', to use of a mathematical
expression. Although every being is determined here in-itself, it is, at the same time, the expression
of a particular activity to be performed. And in the perception of this activity, a fundamentally new
direction of seeing opens up for the human being: the perception of 'objective causality'."

[90] E. Cassirer, "Form and Technology", p. 46 (ECW 17, p. 178): "This results from that original
relation in which all artistic beauty stands in relation to the grounding and original phenomenon of
expression."

[91] E. Cassirer, "Form and Technology", p. 46 (ECW 17, p. 179): "[…] a quite different connection
between the creator and his work prevails in the purely technological sphere as compared to the
artist and his work. The completed object, in becoming actual, belongs to reality. It is situated in a
pure world of things whose laws it obeys and by whose measure it wants to be measured. It must
henceforth speak for itself, and it speaks only of itself and not of the creator to whom it originally
belonged."

deconsecrated part of ritual which would have ended up becoming autonomous from it.[92] Of course, this remark only holds in an interpretive framework within which the difference between "real" and "magical" is clearly marked, which is not the case with ritual when it is actively experienced. But this enables Cassirer not so much to describe the real itinerary which would have lead to the progressive separation between technology and ritual, but rather to better grasp their opposition: whereas ritual is still geared towards the magical possession of the world through the effect of a desire, technology results from the necessary acknowledgement of *renouncing* the possession of what would have value in the world on the sole basis of desire.[93] It is precisely due to this renouncement that technological expression is linked to the same type of issues as science: whereas its expression should be entrenched upon its past, it is nevertheless inasmuch as the technological object is confronted with a plurivocity of senses governed by a same norm that it is transformed from within.

2   Evocation

As seen above in the introduction to this chapter, all forms possess an *internal coherence* which ensures their persistence and a *plasticity* ensuring their adaptation. Until now, it has mostly been question of the way in which *idiomatic* expression participates to the internal coherence of the form—its signification—and it must now be shown how its plasticity—its relation to sense—also depends on it. This is the role attributed to the transcategorical operator called "evocation", following K. Bühler, and of which the denomination has varied over the course of Cassirer's career. Its function consists in operating the possibility of a transfer of signification from its always localized idiomatic expression. Hence, it could also be called "propagation" or "diffusion". It is therefore a matter of seeing how this evocation or propagation operates in the four symbolic forms which we have elected to study.

---

[92]E. Cassirer, *The Philosophy of Symbolic Forms*, vol. 2, pp. 182–183 (ECW 12, p. 214): "Hunting is no mere technique for tracking and killing game; its success does not depend merely on the observance of certain practical rules but rather presupposed a magical relation which the man creates between himself and his quarry. It has been observed among all North American Indians that the "real" hunt must be preceded by a magical hunt which sometimes lasts whole days and weeks and which is bound up with very definite precautionary measures, with all sorts of taboo regulations."

[93]E. Cassirer, "Form and technology", p. 29 (ECW 17, p. 156): "Because the will jumps directly towards its goal in the magical identification of 'I' and 'world', no true mutual determination between them occurs. For every such confrontation calls for proximity as well as distance, empowerment as well as relinquishment, the force of grasping but also the force of keeping something remote. It is precisely this double process revealed in technological activity that differentiates it from magical activity."

2.1 Mythical Evocation

In order to clarify the nature of what Cassirer means by "evocation" in the case of
myth and of magic which are, in his view, part of the same set, it would be appropriate
to begin by taking a methodological precaution. Traditional explanations attempting
to justify the existence of myths and of magic generally base themselves on an
implication which Cassirer considers to be erroneous: myths and magic would be the
outwards projection of humanity's *internal* desires; they would then be the expression
of a *primitive* state of humanity having been *superseded* because the power of desire
is no longer the sole force imprinting upon human history its direction. For Cassirer,
the distinction between inside and outside does not have an absolute value but is only
relative to the interpretational framework within which it is inscribed and therefore
supposes that it had previously been justified. A mode of thinking of which the
frameworks of reference are not our own should not be dismissed as "primitive" (in
the sense of being *retrograde*) as long as we have not uncovered its basic coherence.[94]
Once this reference framework has been placed within the open system of symbolic
forms as described by Cassirer, it then may be possible to consider it as "primitive"
in the sense of being anterior, all the while seeing in the notion of projection a simple
artifact coming form another interpretational framework.[95]

2.1.1 The Mythical Evocation of Space, Time, and Number

It is now time to specify the nature of the semiotic means at the service of the displacement of signification and of its opening to sense.[96] The question is therefore
that of knowing how mythical signification, having appeared by condensation of an

---

[94]E. Cassirer, "Form and Technology", pp. 25–26 (ECW 17, pp. 152–153): "The worldview of *homo
divinans* is supposed to come about through the projection of his condition onto reality; he sees in
the external world what is going on within himself. Inner processes that take place entirely within
the soul are transferred outside of the human body. Drives and wilful movements are interpreted
as strengths that intervene directly into events, steering and altering them. However, from a purely
logical perspective this explanation is marred by a petition principia – it confuses that which is to
be explained and the ground of explanation. When we reproach indigenous peoples for 'confusing'
the objective and subjective, for letting the borders of both areas flow into one another, we are
speaking from the standpoint of our theoretical observation of the world founded on the principle
of 'cause', on the category of causality as the condition of experience and the objects of experience.
These borders are not 'in themselves' objectively before us; rather, they must first be set down and
secured, they must first be erected by mental labour. The manner of setting these borders take place
differently according to the direction in which it moves."

[95]Cf. supra § 33.

[96]E. Cassirer, *The Philosophy of Symbolic Forms*, vol. 2, p. 79 (ECW 12, pp. 94–95): "[...] the
isolation of the immediate datum is overcome; [...] we must seek to understand how particulars
are woven into a whole. [...]. And [...] the concrete expressions of this wholeness, its intuitive
schemata, prove to be the fundamental forms of space, time, and finally of number in which the
factors which appear separate in space and time, the factor of "coexistence" and the factor of
"succession", permeate each other."

emotion, becomes transposable to multiple situations and thereby acquires a certain
semantic perimeter which goes beyond the here and the now and which is susceptible
of modifying itself according to the own finality of desire.[97] Cassirer responds to this
by putting into play the relations between different symbolic forms: his response,
being oriented as we shall see towards the relation between myth and language in his
1925 text *Language and Myth*, has more to do with the relation between myth and
scientific knowledge in the volume devoted to myth in *The Philosophy of Symbolic
Forms* published the year before. The two perspectives do not completely overlap
and the one deployed in *The Philosophy of Symbolic Forms* is related to this first
presentation of the three primordial symbolic forms of which he tries to present
the mutual relations according to modalities which he maintains as close as possible to science.[98] Cassirer defines in this text the aforementioned semantic perimeter
as the set of "categories" of mythical thought and distinguishes among them three
fundamental ones: space, time, and number.[99] From this viewpoint, he describes
for example mythical space as being located in an "intermediary" position between
the space of sensible perception and the space of geometric intuition, this notion of
"intermediary" making sense only because Cassirer still posits philosophical understanding in narrow proximity with scientific knowledge and less so in the free play
of the comparison between symbolic forms, as will later be the case. Regardless
of this point which concerns the evolution of the architecture of the philosophy of
symbolic forms, it is most necessary to stress that based on the foundational distinction between sacred and profane,[100] the notion of mythical space enables a general
ordering of the world[101] according to two crucial operations which Cassirer calls
qualification aiming to define a relation of motivation between a place and the thing
which occupies it and the operation of *systematization* inscribing this thing-to-place
relation in a perspective where the motivation extends to the cosmic totality, as in the

---

[97] E. Cassirer, *Language and Myth: A Contribution to the Problem of of the Names of of Gods*, p. 163
(ECW 16, pp. 260–261): "The mode of this concentration, however, always depends here upon the
direction of interest; it essentially depends not upon the content of intuition but upon de teleological
perspective under which it is configured."

[98] Cf. Chap. 3, § 31, in which the ambivalence of the place of language is discussed.

[99] E. Cassirer, *The Philosophy of Symbolic Forms*, vol. 2, p. 80 (ECW 12, p. 96): "Mythical thinking
reveals the same process of schematization; as it progresses it, too, discloses an increasing endeavor
to articulate all substance in a common spatial order and all happenings in a common order of time
and destiny. This striving found its highest mythical fulfillment in the structure of the astrological
world view; […]."

[100] E. Cassirer, *The Philosophy of Symbolic Forms*, vol. 2, p. 85 (ECW 12, p. 100): "In contrast to
the homogeneity which prevails in the conceptual space of geometry every position and direction
in mythical space is endowed as it were with a particular *accent* – and this accent always goes back
to the fundamental mythical accent, the division between the sacred and the profance."

[101] E. Cassirer, *The Philosophy of Symbolic Forms*, vol. 2, p. 86 (ECW 12, pp. 101–102): "But the
immense complexity which results from this, the weaving of all individual, social, spiritual, and
physical-economic reality into the most diverse relations of totemic kinship, becomes relatively
transparent as soon as mythical thinking begins to give it a spatial expression."

case of the astrological system.[102] Time follows a similar perspective: Cassirer makes
use of Usener's remark according to which, in Latin, the term *tempus* referring to
time derives from the term *templum* which designates a separate and already sacred
space but which first signifies a portion of the sky and, by extension, a moment of
the day.[103] In conformity with the properties of qualification and of systematization,
mythical time therefore also ends up extending towards the totality of an order by
distinguishing periods differing in terms of sacredness,[104] without there being a clear
distinction within them of instances from the past, present, and future. Finally, the
case of number also pertains to qualification and systematization: each number possesses its intrinsic sacred feature (the first three numbers, their immediate successor,
the number four, and the number seven, for example[105]) which is not only conceived
as an ordinal but each number, in its own magical effectivity, can penetrate the essence
of all things and thereby put the most heterogeneous of realities into relation.[106]

Cassirer has also studied mythical evocation by comparing the symbolic forms
of myth and of language and by establishing between them a kinship of processes
which we will now examine.

2.1.2 The Kinship Between Evocation in Myth and in Language

As is often the case with Cassirer, it is by playing a form against others that he
manages to specify their relations of proximity or distance: this is particularly the
case with mythical narratives in which the immediately "mythifying" aspect of language can be sensed in what rhetoric calls *paronymy*, that is, the establishment of
solely phonetic relations between words having no true relation in terms of signification. Cassirer recognizes, in this respect, the value of the research conducted by
the historian of religion Max Müller who saw in paronymy the key of the relation
between language and myth: the purely phonetic kinship established between two
words which do not have the same signification then appears to be a capital resource

---

[102] E. Cassirer, *The Philosophy of Symbolic Forms*, vol. 2, p. 93 (ECW 12, p. 109): "In all this we
see the two fundamental features of the mythical feeling of space – the thorough qualification and
particularization from which it starts and the systematization toward which it nevertheless strives."

[103] E. Cassirer, *The Philosophy of Symbolic Forms*, vol. 2, p. 107 (ECW 12, p. 126): "The Latin
tempus, to which corresponds the Greek *temnos* and **tempos* (preserved in the plural *tempea*), grew
out of the idea and designation of the templum."

[104] E. Cassirer, *The Philosophy of Symbolic Forms*, vol. 2, p. 108 (ECW 12, p. 127): "Thus, time
as a whole is divided by certain boundaries akin to musical bars? But at first its "beats" are not
measured or counted but immediately felt."

[105] E. Cassirer, *The Philosophy of Symbolic Forms*, vol. 2, pp. 146–147 (ECW 12, p. 174): "Where
north, south, east, and west are distinguished as the cardinal points of the world, this specific
distinction usually serves as a model and prototype for all articulation of the world and the world
process. Four now becomes the sacred number par excellence, for in it is expressed precisely this
relation between every particular reality and the fundamental form of the universe."

[106] E. Cassirer, *The Philosophy of Symbolic Forms*, vol. 2, p. 143 (ECW 12, p. 170): "Not only
number as a whole but every particular number is, as it were, surrounded by an aura of magic,
which communicates itself to everything connected with it, however seemingly irrelevant."

for the emotional power of language in its most sensible aspect, a power which immediately gives it a mythical resonance. The mythical narrative then appears as a *staging* of paronymy as it expresses the originative force of expression capable of operating an assimilation between syllables on a purely verbal basis. Cassirer, as we have seen, refers on several occasions[107] to an example borrowed from Max Müller concerning one of the Greek creation myths: after having been saved by Zeus from the flood having caused the human race to disappear, an oracle made the prediction to Deucalion and Pyrrha that they would need to rid themselves of the "bones of their mother" in order to ensure the rebirth of the human race; Deucalion then removed the stones from a field and cast them behind: he then witnessed with Pyrrha the birth of a new humanity.[108] For Max Müller, the story remains incomprehensible for as long as the homophony between the Greek words *laooi* and *laas*, "man" and "stone", goes unnoticed: "sowing stones" becomes synonymous with "sowing men", the notion of sowing seeds extending, for no reason other than a phonological one, to minerals and humans. Cassirer refers to the main lines of this analysis[109]: it is the idiomatic and contingent character of Greek phonology which makes possible, for speakers of the Greek language at a particular but non-specified point of history, an *overdetermination* which makes use of a phonological similarity between the two words as a vector for a possible extension of signification (here, sowing seeds) beyond the limits of its original semantic perimeter. Such a "mythological" extension effectively shows that language is not the locus of a simple recording of permanently stabilized significations but rather stems from a continuous activity of sense-shaping. This shaping activity is founded on the tendency towards the unconditioned which Cassirer has shown to concern the propensity of any form to present

---

[107]In E. Cassirer, *Language and Myth: A Contribution to the Problem of the Names of Gods*, p. 133 (ECW 16, p. 230); *The Philosophy of Symbolic Forms*, vol. 2, p. 39 (ECW 12, p. 27); The Myth of the State, p. 19 (ECW 25, p. 22).

[108]It must be noted however that the elements of isotopy which participated in the "decor" set by the myth's scene (the relation between the bones and the stones, the narrative elements setting the narrative's goal as the renewal of the human race, or the analogy between the gesture of sowing seeds and that of casting stones) and which make possible the staging of the paronymy itself are strangely neglected by Max Müller and by Cassirer, while they are capital for the paronymic relation to appear and to produce an effect.

[109]It must however be noted in what respect Max Müller's example may today appear limited. If we keep with what Cassirer indeed said about the matter, Max Müller attributes to paronymy a status of foundation in the formation of all mythologies while he does not recognize the fact that the example he uses to support his thesis *describes less the genealogy of mythology than that of the consonantal alphabet*. The process of construction of the paronymy (the '*la*' in '*la-as*' and in '*la-ooi*') indeed resembles the construction of a word from a root following a process of rebus writing as occurs in all languages using a writing system based on a consonantal alphabet (Phoenician, Hebrew, and Arabic, for example). The paronymy would then, in this case, find itself extended to any formation of words using a consonantal alphabet. Regardless of the extension one wishes to give to the notion of paronymy, it nevertheless remains that it is not impossible that the debates of his time regarding the age of Semitic languages with respect to Indo-European Languages came into play without his being aware.

itself as unique and absolute,[110] that is, as enduring in its existence regardless of its environment, be it to the detriment of other forms. Here, it is the notion of seed sowing which tends to endure in a signification well beyond its original scope by a double extension directed towards minerals and humans, of which the myth says, based on a logic which is its own, that they correspond to one another in terms of sense *because* they are phonologically identical owing to the phoneme/*la*/. Even if Cassirer does not mention it, it is as if the following phonemes in the two words, /*-as*/and/*-ooi*/, could behave towards one another as the two reversible faces of a same pattern, the first playing the role of background and the second playing the role of figure,[111] as if they were the two faces of a same form: from this would emerge original and unpredicted sense deployed through the mythical narrative of Deucalion and Pyrrha. The occurrence of the perception of paronymy is therefore as much an activity of linguistic signification as it is mythical narration through the intermediary of sense.

The example analyzed by Cassirer shows the kinship which can reside between the symbolic forms of language and of myth. They already give an indication regarding the resources of evocation when it is studied in the case of language alone, as we will now see.

2.2 Linguistic Evocation

Cassirer treats the case of linguistic evocation as he does that of mythical evocation, by varying the relations between symbolic forms in order to attempt to approach the notion he seeks to describe. As before, the perspective adopted in *The Philosophy of Symbolic Forms* tends to present the relations of kinship and of distance between the three first symbolic forms on a mode which narrows the gap between philosophical understanding and scientific knowledge. This will no longer be the case later on when the free play of the relations of kinship and distance between symbolic forms will be marked by a greater autonomy of philosophical understanding with respect to any form of knowledge. At this point, the question of evocation in language should be addressed while specifying the types of relations between symbolic forms which Cassirer chooses to favor.

---

[110]E. Cassirer, *The Philosophy of Symbolic Forms*, vol. 1, p. 81 (ECW 11, pp. 10–11): "[…] in the course of its development every basic cultural form tends to represent itself not as a part but as the whole, laying claim to an absolute and not merely relative validity, not contenting itself with its special sphere, but seeking to imprint its own characteristic stamp on the whole realm of being and the wholelife of the spirit."

[111]As in the famous examples of Rubin's vase or the "duck-rabbit".

2.2.1   The Relation, Between Induction and Exemplarity

When studying linguistic evocation assisted by the relation to scientific knowledge, it is upon the notion of relation that Cassirer's analysis mainly focuses because that is where he finds a kinship between the two forms. This kinship is complex however and requires analysis.

The Inductive Quest for Generality

Cassirer's argument concerning the concept of relation is difficult to follow in that it uses rather classical materials—particularly from Aristotle—and, without explicitly stating so, he places them within another context, that of a form of anthropological and linguistic evolutionism which prevailed during the nineteenth century, and which distinguishes more or less abstract "stages"[112] in the elaboration of concepts using language.[113] Cassirer indeed uses, in the context which is his own, a theoretical schema which goes back to the *Posterior Analytics*[114] and which aims to explicate the possibility of abstract relations of a logical nature: the "reference to" would evolve from a simply empirical relation to passively perceived objects and to a fully conceptual relation, this within a ternary inductive schema in which the progressive transition is made from the "sensible" to the "conceptual", transiting through the "intuitive". It should be noted Cassirer adds a fourth stage to these three classical stages, that of the relation itself, which had not traditionally been thematized as such but which the logic of his time did highlight. But furthermore, this evolution, from the "sensible" to "conceptual" abstraction in linguistic expression can also be found within each moment along different modalities.

In what he calls the expression of sensibility, Cassirer distinguishes three stages which he calls "mimetic", "analogical", and "symbolic".[115] The "mimetic" stage aims to describe a state in which the gesture has not yet been distinguished from speech and in which speech itself is conceived as a gesture.[116] The "analogical"

---

[112]This is how Cassirer qualifies the progressive increase in abstraction in *The Philosophy of Symbolic Forms*.

[113]E. Cassirer, *The Philosophy of Symbolic Forms*, vol. 1, p. 289 (ECW 11, p. 262): "So far to be sure, we have only set up an abstract schema of linguistic concept formation; we have outlined its framework, as it were, without entering into the details of the picture. To gain closer understanding of the process, we must follow the manner in which language progresses from a purely «qualitative» to a «generalizing» view, from the sensuous and concrete to the generic and universal."

[114]Cf. Aristotle, *Posterior Analytics*, for example II, 13, 97b 5–10 and II, 19, 99b 25–35.

[115]E. Cassirer, *The Philosophy of Symbolic Forms*, vol. 1, p. 190 (ECW 11, p. 137): "In general, language can be shown to have passed through three stages in maturing to its specific form, in achieving its inner freedom. In calling these the imetic, the analogical, and the truly symbolic stage, we are for the present merely setting up an abstract schema – but this schema will take on concrete content when we see that it represents a functional law of linguistic growth, which has its specific and characteristic counterpart in other fields such as art and cognition."

[116]E. Cassirer, *The Philosophy of Symbolic Forms*, vol. 1, p. 184 (ECW 11, p. 130): "Even today, among primitive peoples, the language of gestures not only continues to exist side by side with

stage begins, in language, with the mastery of sound: precisely because it is not "mimetic" and because it is volatile, it no longer depends on a simple imitation of an external reality.[117] The "symbolic stage"—the term does not have the general scope here it has elsewhere in Cassirer's work[118]—makes possible the sentence in that it manifests signification itself: the sentence in its constitution no longer bears any relation of imitation with an external reality, thing, or gesture, and aims to express signification for itself.[119]

---

the language of words, but still decisively affects its formation. Everywhere we find this characteristic permeation, on consequence of which the "verbal concepts" of these languages cannot be fully understood unless they are considered at the same time as mimetic and "manual concepts" [expression in English]. The hands are so closely bound up with the intellect that they seem to form a part of it."

[117]E. Cassirer, *The Philosophy of Symbolic Forms*, vol. 1, p. (ECW 11, p. 131): "For this dynamic the language of gestures, which is restricted to the medium of space and thus can designate motion only by dividing it into particular and discrete spatial forms, has no adequate organ. In the language of words, however, the particular, discrete elements enters into a new relation with speech as a whole. Here the elements exists only in so far as it is constantly regenerated: its content is gathered up into the act of its production."

[118]One cannot fail to be surprised by the fact that Cassirer uses here the term "symbolic" to designate the terminal stage of the ternary schema, suggesting that this last stage is indeed that which summarizes by itself the whole preceding process—and that it is possible to think that it abolishes the ulterior stages in a way. This probably comes from the fact that the term "symbolic" as it is used here refers to the usage K. Bühler makes of it when drawing a distinction between the "deictic field" and the "symbolic field"; cf. For example, Karl Bühler, *Theory of Language; The representational function of language*, transl. by D. F. Goodwin with collaboration from A. Eschbach, John Benjamins Publishing Company, 2011, p. 159: "When Wegener and Burgmann, the pioneers of an adequate doctrine of the deictic signals of language, list the circumstances that can contribute to the determination of the communicative value of phonetic signs in a concrete speech situation, they rightly mention a great deal, for example, even the profession (and the business) of each of the interlocutors to the extent it is known to the other. [...] and p. 171: "[...] the representational implement of language ranks among the *indirect* means of representing, it is a *medial* implement in which certain *intermediaries* play a part as ordering factors. In language [...] a set of medial factors, the intermediaries of language (to repeat the expression) stand between the acoustic material and the world: in our language, for example, the Indo-European case system is such an implement [mediating between the world of what is represented and the sounds that do the representing]." Bühler's book is posterior to *The Philosophy of Symbolic Forms* but Cassirer was aware of his previous works and he also explains the fact that he did not cite him in *The Philosophy of Symbolic Forms* by invoking editorial explanations.

[119]E. Cassirer, *The Philosophy of Symbolic Forms*, vol. 1, p. 196 (ECW 11, p. 146): "The purely formal accomplishment of reduplication becomes even more evident where it passes from the sphere

Within intuitive expression, Cassirer distinguishes internal intuition corresponding to the expression of space, of time, and of the concept of number[120] from external
intuition corresponding to the expression of the concept of subject (through grammatical distinctions between things and persons[121]) and of possession (by means of
possessive and then personal pronouns[122]). It must be noted that it is particularly the
analysis of intuitive expression which occupies Cassirer in *The Philosophy of Symbolic Forms* following here one of the features specific to the Kantian perspective
which saw in the intuitive framework the possibility for the schematization necessary to the elaboration of scientific knowledge. The stages of the construction of
these concepts stem indeed from an ever-increasingly advanced *functionalization*
(in particular, outside of the realm of one's own body[123]), according to the schema
which Cassirer put into place starting in *Substance and Function*.[124] In conceptual
expression, Cassirer describes the "qualifying" construction of concepts which does
not seek the greatest extension, but the greatest level of determination,[125] in the form
of an arrangement into series.[126] Finally, regarding the expression of relations, he
addresses the problem of the transition to a fully logical and not solely linguistic
representation of relations by means of the usage of suffixes.[127]

---

of quantitative expression to that of pure relation. It then determines not so much the signification
of the word as its general grammatical category."

[120] E. Cassirer, *The Philosophy of Symbolic Forms*, vol. 1, p. 198 (ECW 11, p. 147): "[...] it is only
through the medium of these forms, through the intuitions of space, time and number that language
can perform its essential logical operation: the forming of impressions into representations."

[121] E. Cassirer, *The Philosophy of Symbolic Forms*, vol. 1, p. 253 (ECW 11, p. 216): "In almost all
languages which divide nouns into specific classes, a personal class and an object class are clearly
distinguished."

[122] E. Cassirer, *The Philosophy of Symbolic Forms*, vol. 1, p. 260 (ECW 11, p. 226): "The presupposition of Humboldt is confirmed by the manner in which language expresses personal relations,
not at first by using true personal pronouns, but by means of possessive pronouns."

[123] E. Cassirer, *The Philosophy of Symbolic Forms*, vol. 1, p. 207 (ECW 11, p. 159): "Now the
terms are no longer drawn exclusively from man's own body; but the method by which language
represents spatial relations has remained the same."

[124] E. Cassirer, *The Philosophy of Symbolic Forms*, vol. 1, p. 212 (ECW 11, p. 165): "Here again
we see how tenaciously language clings to the spatial datum as soon as it undertakes to represent
motion and pure activity. As man turns his attention to activity and apprehends it as such, he must
transform the purely objective, substantial unity of space into a dynamic-functional unity; he must,
as it were, construct space as a totality of the directions of action."

[125] E. Cassirer, *The Philosophy of Symbolic Forms*, vol. 1, p. 288 (ECW 11, p. 260): "The words of
language are not reflections of stable concretions in nature and in the perceptual world, they rather
indicate directions which the process of determination may follow."

[126] E. Cassirer, *The Philosophy of Symbolic Forms*, vol. 1, pp. 292–293 (ECW 11, pp. 266–267): "It
complements the special signification of each word as such by adding a new determining element,
which discloses its relation to other words. [...]. For the logical theory of concepts clearly demonstrates that the "serial concept" is not posterior to the "generic concept" in force and significance,
but is an integral part of the generic concept."

[127] E. Cassirer, *The Philosophy of Symbolic Forms*, vol. 1, p. 307 (ECW 11, p. 285): "And it was
this use of suffixes which prepared the way for the designation of pure concepts of relation. What

There would then be something of a harmonious congruence between the inductive movement within a particular sentence of the linguistic expression and the sequencing of these sentences within the totality of the language form because this movement can also be found in the later stages. Even if Cassirer is cautious in stressing that this inductive schema is to be conceived in a methodological and not chronological manner,[128] it does nevertheless remains conceived as enabling a general classification of linguistic phenomena based on the notion of relation which completes it.[129] Furthermore, Cassirer—at least in the beginning of his career—pointed out that this inductive schema *also holds for symbolic forms other than language*, such as science and art which are meant to follow, in their own form, the same evolution from the more "sensible" to the more "conceptual".

Transforming the Articulation Between Signification and Meaning

However, when examined more closely, *this schema is subverted by Cassirer himself* when he addresses the conceptual stage of the language form: reacting from a critical stance to the works of his time in the field of logic, in particular to those of R. H. Lotze,[130] Cassirer points out regarding the example of color that the relation of subsumption between genus and species, typically used as an example for inductive relations enabling an increase in generality, is actually *not the most primordial* when it is a matter of the construction of signification in the sphere of language.[131]

---

first served as a special thing-indication developed into the expression of categorical determination, e.g., of an attributive concept as such."

[128] E. Cassirer, *The Philosophy of Symbolic Forms*, vol. 1, p. 295 (ECW 11, p. 270): "We may attempt to arrange these by taking as our guiding principle the constant progress from the "concrete" to the "abstract" which determined the general development of language; yet we must bear in mind that we have to do not with a temporal, but with a methodological stratification and that in a given historical configuration, the strata which we shall attempt to differentiate may exist side by side or may be intermingled in a variety of ways."

[129] E. Cassirer, *The Philosophy of Symbolic Forms*, vol. 1, p. 318 (ECW 11, p. 299): "In the universal term of relation, the copula, we thus find confirmed the same fundamental tendency of language that we have followed in the linguistic configuration of the special terms of relation."

[130] E. Cassirer, *The Philosophy of Symbolic Forms*, vol. 1, p. 283 (ECW 11, pp. 254–255): "Traditional logicians are convinced that the concept must be oriented purely towards universality and that its ultimate achievement must be to provide universal representations; but then it develops that this essentially uniform striving for universality cannot everywhere be fulfilled in the same way. Consequently, we must distinguish two forms of universal: in one, the universal is given only implicitly, as it were, in the form of a relation disclosed by the particular content, while in the other it emerges also explicitly after the manner of an independent intuitive representation. But from here only one step is required to reverse our viewpoint: to look upon the enduring relation as the true content and logical foundation of the concept, and to oregard the "universal representation" as a psychological accident which is not always desirable or attainable." Cassirer remains here within the strictest of Kantian orthodoxies which, starting in 1762 with *The False Subtlety of the Four Syllogistic Figures Proved*, criticized the logicists of his time for placing abstraction before the relation.

[131] E. Cassirer, *The Philosophy of Symbolic Forms*, vol. 1, pp. 283–284 (ECW 11, p. 255): "Blue and yellow are not particulars subordinated to the genus "color in general"; on the contrary, color "as

As we have seen on several occasions, it is the both qualifying and exemplary construction of an idiomatic expression which is first and which gives itself to be seen in the example of the perception of color. A color indeed does not stem from a relation of generality as manifested in the difference between genus and species because there is no "generality" of a color: it is not "blue" or "yellow" in general which is perceived as a color, but rather such instance of blueness or such instance of yellowness. Why is it then that two colored qualities will be associated within a "same" color? On what rests their belongingness to the "same" color? Cassirer's answer consists in saying that what we call a color already comprises within itself the possibility for serialization, typically in accordance with a functional rather than a substantive relation, in the sense of the difference between substance and function which he brought to light on several occasions. This last remark is capital: for Cassirer, the functional notion of serialization thus appears not only with the linguistic construction of perception, but also as *extending beyond* this form since we can see in it the rise of a truly conceptual determination: the notion of series already comprises a relation of *sense* of a logical nature and not only a relation of signification of a linguistic and perceptual nature. Thence, the functional point of view, present already at the origin of the linguistico-perceptual form but of which the presence appears only once the maximal expansion of this very form has been reached, is a relation which *articulates signification and sense*[132] even if it is not the only one to do so.

We can now draw all the consequences from this construction regarding the nature of the articulation in question. At first glance, Cassirer's exposition remains in a way limited to a *dogmatic* relation to signification, internal to a particular symbolic form, and it is the inductive quest for generality which then manifests in keeping with the principle according to which a symbolic form always tends to occupy all the space of signification in its own order.[133] However, from the standpoint of a *critical* relation between signification and sense, that is, from the point of view of the potential for symbolic transformation specific to each form—when Cassirer likens for example the notion of serialization already present at the moment of the perception of color to

---

such" is contained nowhere else but in them and in the totality of other possible color gradations, and is thinkable only as this aggregate in its graduated order. Thus universal logic points to a distinction which also runs through the whole formation of linguistic concepts. Before language can proceed to the generalizing and subsuming form of the concept, it requires another, purely *qualifying* type of concept formation. Here a thing is not named from the standpoint of the genus to which it belongs, but on the basis of some particular *property* which is apprehended in a total intuitive content. The work of the spirit does not consist in subordinating the content to another content, but in distinguishing it as a concrete, undifferentiated whole by stressing a specific, characteristic factor in it and focusing attention on this factor. The possibility of "giving a name" rests on this concentration of the mind's eye: the new imprint of thinking upon the content is the necessary condition for its designation in language."

[132] We could almost speak here of "sublation" (*Aufhebung*) in the Hegelian sense, except that it is a matter here of a suprasumption which is by no means necessary: The relational perspective *is itself but a qualifying and exemplary construction among others* and it therefore has no absolute necessity.

[133] Cf. above, Chap. 3, § 2.

the logical notion of relation—we can view the articulation of signification and sense in a whole different manner, as a *permanent transformation* of this articulation itself, according to modalities which are not anticipated beforehand and which are therefore neither circumscribed to an inductive schema of the type which may be described from within a particular symbolic form, nor to a strictly functional perspective. *What Cassirer understands as such by "symbolic" would then be the manifestation of this double movement:* on the one hand, a movement of expansion of the signification internal to a form—analogous to a growth and describable in the classical terms of the inductive schema aiming the greatest generality—and, on the other hand, an outwardly oriented movement which is specific to sense and which by no means seeks generality, but which rather seeks the possible relation to other forms based on a qualifying and exemplary construction. It is this double movement which clarifies the nature of what has earlier been called[134] a "transcategorical operation", apt to transform the contents but also the interpretational frameworks in which the contents become significant. From a linguistic standpoint, the notion of metaphor makes this double movement fully manifest.

2.2.2 The Innovative Metaphor at the Core of Linguistic Evocation

The question of the status of the metaphor seems capital for Cassirer because it is in it that lies the possibility for a recomposition of the relations between signification and sense which is not only oriented towards functionalization. We have already encountered in the example of Deucalion and Pyrrha borrowed from Max Müller the way in which signification attracts sense by playing with the relation between background and form as if they were the two faces of a same reality. Even if the analysis was more centered upon the relations between language and myth, the relation was indeed a metaphorical one, and it should now be characterized from a more specifically linguistic perspective.

For Cassirer, metaphors have two acceptations which seem to intersect with the Kantian distinction between analytic judgment and synthetic judgment. The first designates, in a classical manner, the transfer of signification from one already constituted domain towards another already constituted domain,[135] as in the example mentioned earlier in which the notion of victory is supposed to transition from the dance performed by the group of women to the battle fought by the group of men. The second acceptation, which Cassirer calls "radical metaphor",[136] designates the way

---

[134]Cf. Chap. 4, § 2.

[135]E. Cassirer, *Language and Myth: A Contribution to the Problem of the Names of Gods*, p. 204 (ECW 16, p. 302): "This use of metaphor, however, obviously assumes the sensory content of individual formations, as well as their linguistic correlates, as already *given*, as fixed quantities; only after these elements have been linguistically determined and fixed as such can they be exchanged for one another."

[136]E. Cassirer, *Language and Myth: A Contribution to the Problem of the Names of of Gods*, p. 204 (ECW 16, p. 302): "This transposition and exchange, which already exchanges the vocabulary of language with its material, must be distinguished from the genuine "radical" metaphor, which is a

in which signification manages to extend beyond the already constituted perimeter of its exercise by creating a new domain of sense. Such is the case in the example of the myth of Deucalion and Pyrrha borrowed from Max Müller and as presented by Cassirer: the narrative draws a likeness between *laas* and *laooi* by identifying a same background /*la*/all the while presenting by means of the narrative two distinct forms /*-as*/and /*-ooi*/as being interchangeable, that is, as alternately playing for one another the role of figure and ground, as Gestalt theory put forth using very famous examples where two spatial configurations become alternatively and reciprocally figure and ground with respect to a same surface constituting a shared background to both. The metaphor in this case is "radical" because it institutes a *new domain* pertaining to the world of signs in which a background bearing the value of indistinct "matter" distinguishes itself from a form bearing a determined "signification", both values being interchangeable.

Perceiving a paronymy is therefore as much an activity of linguistic significa-tion as of mythical narration and we understand why Cassirer sees in the "radical" metaphor their common source: attributing a sound to an emotional drive in order to form a linguistic sign is *already* a radical metaphor inasmuch as this operation of symbolic pregnancy conjointly produces the experience of realities which are com-pletely heterogeneous with respect to one another, as it takes a simple impression of everyday life to place it within the sphere of the sacred in the case of the "pri-mary mythical configuration".[137] We thus discover that "symbolic pregnancy" has the "radical metaphor" for foundation and that it is the latter which brings within a same movement sense and the sacred into existence.

Without Cassirer being very clear regarding the way in which to articulate the two acceptations of the term 'metaphor', these appear to have been conceived as genetic *stages* of a single process: it is the trans-domain flexibility of sense enduring through various contexts which manifests the presence of an investment of sense in a physical medium of any sort, an investment which Cassirer names "fundamental metaphor". Cassirer uses the example of the morpheme *–st* which, in various Indo-European languages and since millennia, signifies the fact of remaining stable[138]: a result of the "fundamental metaphor" which invests a phonological medium in order to express stability, the morpheme *–st* is also metaphorical in the second acceptation,

---

condition of the formation of language as well as the mythical formation of concepts itself.[…]. What takes place here is nt simply a transfer; rather, it is a genuine metabasis eis allo genos; it is not simply a transition from one already existing genus to another but the very creation of the genus itself to which the transition proceeds."

[137] The expression is found on p. 205 in E. Cassirer, *Language and Myth: A Contribution to the Problem of the Names of Gods* (ECW 16, p. 303).

[138] E. Cassirer, *The Philosophy of Symbolic Forms*, vol. 1, pp. 191–192 (ECW 11, p. 139) subscribes in particular to the point of view of G. Curtius whose *Grundzüge der griechischen Etymologie* he cites: "All the peoples of our family from the Ganges to the Atlantic designate the notion of standing by the phonetic group *sta*; in all of them the notion of flowing is linked with the group *plu*, with only slight modifications. This cannot be an accident. Assuredly the same notion has remained associated with the same sounds through all the millennia, because the peoples felt a certain inner connection between the two, i.e., because of an instinct to express this notion by these particular sounds."

inasmuch as the permanence of its value only appears through the intermediary of the various contexts within which it is to be found. The variability of contexts is therefore not that which opposes the stability of sense, but on the contrary that which accomplishes it through a process: the invariant (here, the *–st*) is then defined by its *capacity to lend itself to differentiation*. In the end, this is what characterizes in the most general manner possible the articulation of signification and sense.

2.3 Scientific Evocation

At first glance, the idea that the evocation or propagation of sense may have a semiotic role to play in scientific knowledge appears difficult to conceive, as completely directed as it seems to be towards the most univocal and most permanent determination of signification possible. Yet, evocation intervenes at several levels in the constitution and evolution of scientific knowledge and, from this perspective, if univocal determination is indispensable to the constitution of objectivity, *it is certainly only one of its stages*. As we have indeed seen on several occasions, scientific knowledge bears with other forms of knowledge unstable and often hidden links. It is therefore necessary to now attempt to characterize the notion of evocation in its own right and to examine it in closer detail in the context of scientific knowledge by using a number of examples which we have described earlier.

2.3.1 Scientific Evocation with Respect to a Specific Form of Knowledge

First, two levels of intelligibility should be distinguished when seeking to characterize the role which the evocation of sense could play within a same body of scientific knowledge. Indeed, what Cassirer's analyses have tended to show is that what appears to form a rupture at the level of scientific content or even at the level of the interpretational framework which makes possible a certain type of objects is not necessarily so at the epistemological level: if it is indeed fitting to radically oppose Newtonian and Einsteinian physics not only with respect to the objects they determine but also in what concerns the interpretational frameworks of a geometrical nature they deploy,[139] it is just as fitting to show the profound continuity at the *epistemological* level when interpreting the evolution from the one to the other as an evermore sophisticated *functionalization* of the more general concepts of physics, in this case, the concept of space-time. This continuity is indeed of an exclusively epistemological nature because "space" and "time" do *not have the same signification* in one or in the other physical system, even if they share the same quest for functionalization as *sense*. Two consequences stem from this.

It must first be noted that it is not mathematics as specific knowledge which could establish an epistemological continuity between the two physical systems we have

---

[139]Cf. Chap. 2, § 231.

just mentioned: as shown, for example, by that which the Kantian tradition has called the "Copernican revolution", it is not increasingly complex mathematics enabling to better account for the epicycles in Ptolemy's system which made it possible to formulate the epistemological hypothesis of heliocentrism. Such an epistemological hypothesis involved the attribution of a *new signification* to space, one of a functional type, and this being something that mathematics are only likely to do when adopting a truly epistemological point of view which varies the *sense* of space.[140] Thus, the transition from substantial scientific knowledge of the Ptolemaic type to functional scientific knowledge of a Copernican type, detached from the sensible, is also not the locus where scientific knowledge alone could access the free play which characterizes sense.

It must then be noted that the relations between relatively autonomous theories belonging to a same body of knowledge, theories which sometimes contradict one another as was the case during the age of Einstein with mechanics and electromagnetism, do not take the form of a *reduction* to one amongst them—in this case a reduction of electro-magnetism to mechanics—but that their reciprocal relations transition through a *unification* within a theory likely to contain both, be it at the cost of an integral reinterpretation of the notions of space and time. Then again, it is by means of a new epistemological point of view having meaning for motor that a radical transformation of physics becomes possible.

2.3.2 Scientific Evocation in the Interrelations Between Forms of Knowledge

As was previously the case, scientific evocation plays a role in the relations of reciprocal influence or of distance between forms of knowledge if we situate these relations from an epistemological point of view. We have seen in particular with the case of the emergence of physics as a functional science that such emergence had been preceded during the Renaissance by a revolution in the usage and interpretation of signs regarding topics which had nothing to do with physics. Attempting to establish a *causal* relation between a new usage of signs and the emergence of this new conception would again be an attempt to reduce the relations between forms of knowledge to the sole perspective of a particular form of knowledge—in this case, that of physics—and would lead to a reductionist point of view which has never been that of Cassirer. Quite to the contrary, only an opening to sense enables, for Cassirer, to view the relations between forms of knowledge from a truly epistemological perspective. It is only then that it becomes possible to bring transversal mathematical concepts into play without supposing any unity of signification between the domains considered. Such is the case, according to Cassirer, with the concept of group which

---

[140]E. Cassirer, "Mythic, Aesthetic, and Theoretical Space", in *Ernst Cassirer, The Warburg Years (1919–1933)*, p. 325 (ECW 17, p. 419): "Space does not possess an absolutely given, final, and fixed structure; rather, it acquires this structure only by virtue of the general coherence of meaning within which its very construction is accomplished. The function of meaning is the primary and determining one; the structure of space is a secondary and dependent element."

we have already mentioned,[141] and which appears to him to be, beyond its mathematical signification, usable elsewhere—in particular in the theory of perception. The constancy of perceived objects despite changes in lighting, in angle, and in distance represents a challenge indeed for the theory of perception because we do not see how an object could remain the *same* even if all the physical conditions of its perception *vary*. Cassirer notes that the solution consisting in supposing that individuals unknowingly produce a mathematical calculation which would permanently reestablish the conditions of invariance, in the manner of Helmholtz who spoke of "unconscious inferences", does not further advance the problem's resolution because the hypothesis of unconscious logical implications seems completely *ad hoc*. In fact, for Cassirer, a theory of perception which rests upon the notion of logical inference is of a far too intellectualist nature because it supposes that perception is only possible under the condition that logic informs matter which had previously been reduced to a chaos of impressions. It then becomes inevitable to consider the inferences to be unconscious because they were not conceived by the perceiving subject. In order to avoid such dualism, it is necessary, according to Cassirer, to change *both* the theory of perception and the mathematics which are related to it: only Gestalt theory authorizes a rationalist theory of perception without an intellectualist and dualist bias. As Wertheimer has shown, there are indeed laws of constitution of forms which enable, *at the level of perception itself*, to justify the constancy of objects,[142] and there is a most adequate corresponding mathematical expression accounting for these laws: the concept of invariance under a transformation group.[143] Thus, a new theory (in this case, that of perception) does not consist in reducing a theory to another one deemed to be sounder (here, that of logical inference) by adding the *ad hoc* concept of "unconscious inference" in order to ensure such reduction. The novelty of a theory resides in its capacity to reorganize all the elements present by proposing a new unification within an interpretational framework which respects the data all the while proposing, when appropriate, a possible mathematical schematization.

---

[141]Cf. Chap. 1, § 12.

[142]These laws describe the collective behavior of elements as being both enclosed within a given field (proximity); morphologically similar (similarity); moving in a same manner (common destiny); enabling to follow a straight line (good continuation); more united in a region where there are no gaps (enclosure); presenting a relation between the form and the general values present in the organization of a field (pregnancy).

[143]E. Cassirer "The Concept of Group and the Theory of Perception", *Philosophy and Phenomenological Research*, vol. 5, n° 1, (1944), p. 5 (ECW 24, p. 214): "In the following reflections, I shall attempt to set forth an inner connection – epistemological in nature—between the mathematical concept of group and certain fundamental problems of the psychology of perception as the latter have been more and more distinctly formulated in the last decades. [...]. Our ultimate aim is to bring out clearly a certain type of concepts which has found its clearest expression in abstract creations of modern geometry. But the type in question is not confined to the geometrical domain. It is, on the contrary, of far more general validity and use. The application of concepts of this type extends both farther and deeper. Metaphorically speaking, it extends down to the very roots of perception itself."

2.4 Technological Evocation

The case of technological "evocation" or of its "propagation" has the particularity of being mainly addressed by Cassirer with respect to tools: this supposes that he considers as granted the existence of a continuity between tools and technology,[144] a position which may elicit controversy. However this may be, in Cassirer's view, the tool already contains the elements that technology, and in particular mechanical technology, will only deploy on a different scale. The utilitarian aspect of tools which are used as exterior means manifests no internal necessity inasmuch as it may be possible to do without them. On the other hand, as soon as the tool is itself taken as an object of attention in its own right, it acquires its own autonomy and becomes an instrument of mediation which can develop indefinitely, under the most varied and complex aspects because it has acquired the specific field within which to deploy its action. Thus, technological "evocation" is linked more to the stage during which the tool's own field is constituted, governed by a norm requiring the tool to be always capable of being modified in function of the environment in which it is used.

It then becomes possible to epistemologically compare this symbolic form with other forms, in particular those of science, art, and language. In what concerns the two first forms, the relations are relatively simple to analyze, even if they would deserve developments which Cassirer does not make.

Regarding the relation between science and technology, Cassirer remarks the narrow kinship existing between the two, a kinship he had not emphasized prior to his 1930 article on technology. There lies a novel fact which opens onto an original perspective regarding the birth of science, even if Cassirer remains rather untalkative with respect to the relations between *substantial* technology and science as they may have prevailed during Antiquity or the Middle-Ages. It is rather regarding the modern era that he notes a narrow collusion between science and technology[145] on the one hand, and between technology and Renaissance art on the other hand[146] inasmuch as

---

[144]E. Cassirer, "Form and Technology", p. 30 (ECW 17, p. 158): "To make this clear, we need not look at the complete unfolding and present structure of technology. A basic circumstance presents itself in the most ordinary and inconspicuous phenomena, in the first and simplest beginnings of tule-use, more clearly than in almost all the marvels of modern technology."

[145]E. Cassirer, "Form and Technology", p. 43 (ECW 17, p. 175): "Technological work and theoretical truth share a basic determination in that both are ruled by the demand for a 'correspondence' between thought and reality, an 'adequatio rei et intellectus'.

[146]E. Cassirer, "Form and Technology", pp. 42–43 (ECW 17, p. 174): "For it is no way the 'abstract', pure theoretical knowledge of the laws of nature that leads the way, proving first the technological aspect of the problem and its concrete technological activity. From the very beginning, both processes grasp one another and as it were keep the balance. Historically, this connection can be made clear when we look back at the 'discovery' of nature' that has taken place in European consciousness since the days of the Renaissance. This discovery is in no way the work of only the great researchers of nature – it returns essentially to an impulse originating out of the questions of the great inventors. In a mind like that of Leonardo da Vinci the intertwining of these two basic orientations appears with a classic simplicity and depth. What separates Leonardo from mere bookish learning, from the spirit of the 'letterati', as he himself called it, is the fact that 'theory' and 'praxis', and poiesis, penetrate one another in his person in a completely different measure as never before."

technology is both a human creation which is similar to art in this respect but which works within a world it seeks to make objective in the same manner as does science.

In what concerns the relation to the last form, that of language, a specific difficulty arises due to the fact that technology constitutes, in Cassirer's view, an activity which is *as primordial* as language. It can indeed be conceivable that, in the same way as the action of technology does not only involve using tools but also acting *upon* tools, language too does not only signify by means of sentences, but also acts *upon* sentences and that the indefinite continuation of work *upon* the forms which are language and technology make them both the very paradigms of the instituting power specific to the notion of symbolic form. However, the type of evocation they make possible clearly distinguishes them: whereas language possesses and expressive basis of an idiomatic nature of which the evocative scope extends by functionalization and metaphorization, any tendency towards evocative expression within the framework of technology is rather an indicator that the difference between the subjective and the objective needs to be reinforced in order for its evocative power to be conceived, as much as possible, on the mode of the overcome. This is for example the case with the technology of writing which Cassirer describes as being, at first, essentially magical,[147] but which tends to exceed this stage and to progressively make it possible for signs to acquire their arbitrary nature. From this perspective, the primordial aspect which these two symbolic forms have in common does not account for the reason for which they are nevertheless engaged towards two opposite directions of sense, with language remaining the closest to the mythical ground by which its expression manifests, and with technology tending towards the objectivity of science. It is probably because *voice* as it is used in speech preserves at all times an expressive force by being addressed to a listener, may such a listener be one's own self, and this has no equivalent[148] as regards the evocative expression specific to technology.

---

[147] E. Cassirer, *The Philosophy of Symbolic Forms*, vol. 2, p. 238 (ECW 12, p. 278): "The written sign is not at once *apprehended* as such but is viewed as a part of the objective world, one might say, as an extract of all the forces that are contained in it. All writing begins as a mimetic sign, an image, and at first the image has no significatory, communicative character. It rather replaces and "stands for" the object."

[148] As Cassirer remarks concerning the constitution of deictics such as with the "I"/"you"/"he" distinction, such a distinction only becomes possible from the moment when a situation of inter-locution is constituted into an object proper, a situation in which each speaker modulates his or her speaking in function of that of the others, according to a greater or lesser degree of proximity of his or her body within space. Cf. E. Cassirer, *The Philosophy of Symbolic Forms*, vol. 1, pp. 192–193 (ECW 11, pp. 140–141): "The use of certain differences and gradations of vowels to express specific objective gradations, particularly to designate the greater or lesser distance of an object from the speaker, is a phenomenon occurring in the most diverse language and linguistic groups. Almost always *a, o, u* designate the greater distance, *e* and *I* the lesser. Differences in time interval are also indicated by difference in vowels or by the pitch of vowels. In the same way certain consonants and consonantal groups are used as "natural phonetic metaphors" to which a similar or identical significatory function attaches in nearly all language groups – e.g., with striking regularity the res-onant labials indicate direction toward the speaker and the explosive lingual direction away from the speaker, so that the former appear as a "natural" expression of the "I", the latter of the "Thou"."

3 Objectification

With "objectification", it is question here, in accordance with the conventions which we have adopted at the beginning of this chapter, of the third transcategorical operator which Cassirer also calls "signification".[149] As we have also already noted, even if science appears to be the paradigmatic example of this operator, it is however only an example since any symbolic form must be epistemologically analyzable on the basis of the three operators of which the interplay makes possible the multiform elaboration of meaning. It is for this reason why that which Cassirer means by "objectification" must be also found elsewhere than in science, in accordance with the expression specific to the symbolic form in which objectification manifests itself.

The difficulty specific to the study of the operator of objectification pertains to the nature of the notion of objectivity: on the one hand, it is *constructed*—in particular once the purely arbitrary dimension of the sign has been acquired as well as the related conceptual distinction between that which pertains to the sphere of the subject and to that of the object. On the other hand, *it is not constructed*, inasmuch as objectivity must end up encountering a certain resistance from reality, conceived to be independent from the means which make its access possible. There is therefore something of a paradox here inasmuch as the operator of objectification must enable to account for the *construction* of objectivity interpreted as *non-constructed*. Therefore, it is in fact the conception which Cassirer holds concerning objectivity of which it is question here, a conception that we have already addressed laterally in the two first chapters which concern more directly the exact sciences and the natural sciences but which must be studied here in a general manner in the four symbolic forms we have chosen to study. These remarks will be shorter than those pertaining to the two other operators because the epistemological perspective which constituted our undertaking's angle of approach has already prepared the ground.

3.1 Objectification in Myth

It is assuredly paradoxical that the symbolic form of myth and magic may be associated with the notion of objectification since everything in it seems to oppose a clear distinction between the subject and object upon which the possibility for objectivity is founded. As already seen, the mythical and magical world is populated with forces endowed with volition which transform its appearance, and it is possible for one to change its course while capturing such forces to one's own advantage, which plunges us into what then appears to be a phantasmagorical chaos. And yet, there is indeed in this symbolic form something which pertains to the operator of objectification because it is already question of *explanatory* and *independent* principles describing

---

[149]Cf. Chap. 3, § 32. The choice of the term "representation" comes from the fact that beyond its being inherited from the work of Bühler, it enables to use more specifically the term "signification" in relation to the term "meaning", as shown in Chap. 3, § 12.

a stable order.[150] Such is, for example, the case with the principle of causality that is indeed present in the mythical and magical world[151] but which manifests itself without marking a clear-cut difference between what pertains to the subject's desires and what pertains to the object alone. Negatively, the mythical and magical world is therefore epistemologically defined as that world *which does not have the internal resources* in order to operate the first conceptual distinction between subject and object, a distinction which would enable to use concepts by recognizing them for what they are: signs inscribed within a symbolic relation with the object.[152] However, the fact that it may be possible to exit the mythical and magical world, as Cassirer described regarding the historical case of the crisis having given birth to Greek science, requires that this analysis be completed: conceptual elements such as causality must *indeed be present* in the mythical and magical world since it is possible to leave it. Thus, precisely because they are not the object of attention for their own sake in the world of myth and magic, their presence must be conceived according to the semiotic modality of *obscurity*. Positively, the mythical and magical world have the specificity of being related in a very particular manner with concepts by way of meaning, a relation of which the value must be epistemologically recognized in itself without seeking to reduce it to anything other than itself: the obscurity of which it is question here is therefore constitutive of the symbolic form of myth *as such*.[153] The phantasmagoria of mythical stories and of magical procedures is therefore not a pathology of meaning to be eliminated by all means: it is the direction taken by meaning in the symbolic form specific to myth when concepts of objectivity are not governed by experimental protocols and by the idea of demonstration.

3.2 Objectification in Language

As in the case of myth, it seems at first glance paradoxical to consider language in its objectificational dimension when we know that it is precisely during a crisis with respect to the mythifying powers of language that the concepts of science and

---

[150]E. Cassirer, "Form and Technology", p. 26 (ECW 17, p. 153): "If we assume that the principle of 'causality' and the question concerning the 'ground' of being and the 'causes' of events already prevail in the magical view of nature, the partition between magic and science falls away."

[151]E. Cassirer, "Form and Technology", p. 27 (ECW 17, p. 154): "[magic] grasps nature as a strictly determined sequence of events and seeks to penetrate into the essence of this determination. It knows no coincidence. It rises to the conception of a strict uniformity of events."

[152]E. Cassirer, "Form and Technology", p. 27 (ECW 17, p. 154): "Magic admittedly differs from science in its result but not in its principle and its problem. This is the case because the principle 'like causes, like effects' governs it as well, giving it its generally apparent character. That it is not able to employ this principles in the same sense as the theoretical science of nature is not, according to Frazer, due to a logical reason but only to a factual one. It is 'primitive' not in its form of thought but in the measure and the security of its observation too fluctuating and uncertain, for it to be able to erect truly durable empirical laws. The consciousness, however, of lawfulness as such has been awakened in it and is tightly and steadfastly held onto by it."

[153]Cf. supra § 140.

of philosophy came about. And as is the case with myth, the conceptual elements must nevertheless be present in language, failing which the collective attention could not focus upon them nor work on them for their own sake after the appearance of science. As we have indicated above,[154] language is rich in conceptual potentialities, which it exploits in its own way without however achieving specifically conceptual determination.[155] These conceptual elements therefore pertain, in the symbolic form of language, to the constitutive obscurity of which we spoke earlier concerning myth and which Cassirer described while presenting a certain number of examples.

In the analysis of the operator of objectification, we will revisit in a quite different manner Cassirer's aforecited example[156] concerning the lexicon of numbers because, inasmuch as they appear to be endowed with an independent objectivity of an ideal and necessary nature, they seem to be most removed from any process of construction dependent upon linguistic idiomaticity. We could have indeed imagined that the terms designating numbers aim above all to characterize the cardinality individually defining "abstract objects" but Cassirer notes in reference to the example of Sotho that this is not what occurs: the lexicon does not directly characterize the numbers as if it constituted labels to affix upon independent entities, but rather characterizes the *operations* enabling to construct them by arranging elements in a linear order, a *series*, that is, by performing a count.[157] The way to perform the count refers indeed in the case of Sotho to the body and to its particular way of apprehending the notion of sequence (by distinguishing, in this case, the fingers and the toes) but this particular manner is nevertheless the carrier of a *necessary order* since the counted elements must always be traversed in the same manner.[158] The idiomatic character of the formation of the lexicon thus enables to ensure the coordination of the gestures in shared activities and it is indeed this character which is visible in the lexicon of numbers in Sotho: the name attributed to a number is conceived as an *injunction*

---

[154]Cf. supra § 153.

[155]E. Cassirer, *The Philosophy of Symbolic Forms*, vol. 1, p. 315 (ECW 11, p. 299): "And even where language has progressed to the point of encompassing all these specifications of existence in a universal expression of "being", there remains an appreciable difference between even the most comprehensive expression of mere *existence* and "to be" as a an expression of purely predicative "synthesis"."

[156]Cf. supra § 153.

[157]E. Cassirer, *The Philosophy of Symbolic Forms*, vol. 1, p. 230 (ECW 11, p. 187): "Thus the numerals do not so much designate objective attributes or relations of objects, as embody certain directives for the bodily gesture of counting. They are terms and indices for positions of the hands or fingers, and are often couched in the imperative form of the verb. In Sotho, for example, the word for "five" means literally "complete the hand", that for "six" means "jump", i.e., jump to the other hand. […]. By this method, the motions of arranging the objects are coordinated with certain bodily motions which are conceived as running in a certain order."

[158]E. Cassirer, *The Philosophy of Symbolic Forms,* vol. 1, p. 231 (ECW 11, p. 188): "But one thing is accomplished: a very definite order is observed in passing from one member of a manifold to another, even though this manifold is determined in a purely sensuous way. In the act of counting, one part of the body does not follow another arbitrarily, the right hand follows the left, the foot follows the hand, the neck, breast, shoulder follow the hand and feet in accordance with a schema of succession which is conventional, to be sure, but is in any case strictly observed."

aiming to coordinate the activity in a uniform manner regardless of the person who is counting. What the Sotho language retains is therefore the particular manner of bringing a norm into effect for all speakers by circumscribing in an idiomatic manner their field of activity—in this case, that of counting, but also more generally.[159] We could then think that at first glance, language only accompanies an activity which is already underway: as a sort of spoken supplement of an explanatory or descriptive nature, it would only play a secondary role, whereas the activity it appears to accompany would, for its part, be first and above all invariable. But the example described by Cassirer shows that this is not the case: language, by making possible the decomposition of activities,[160] institutes *a reproducible order* which can use any medium and modify it if needed, as shown with the example of the suffix *-keit*.[161] Thus, the idiomatic expression of language, as random as it may appear, participates to the very elaboration of these activities.

Yet, the example of counting in Sotho could prove misleading if it suggested that the idiomatic expression is actually but a simple supporting element, indeed indispensable to the elaboration of a uniform rule enabling identical reiteration, but having value only inasmuch as it makes possible *something else than itself*, that is, the apprehension of *independent* objects such as numbers. The case of the determination of numbers would then skew the perspective inasmuch as it would tend to focus on the independent aspect specific to the constitution of objectivity. But it is not, in general, the aim conferred to activities because such a point of view could lead to think that the object exists prior to the activity which presides over its construction.

In fact, a linguistic activity does not pursue an aim which would be indexed over a type of exterior object. But it would be a mistake just as well to construe objectivity as deriving only from the collective agreement of a community, as if it would be enough to consider it objective just because several people "previously"— a typically mythical relationship to the passing of time—agreed to carry it out the same way. Objectivity in the linguistic domain depends on the fact that the activity is recognized as being carried out for its own sake, i.e. that *the activity itself is the very object of social concern*. In short, the "object" is not this exterior term which would govern the pursuit of the activity, it is indeed rather *the activity which is the focus of collective attention*. This collective attention directed towards linguistic activity manifests itself as an object *under the aspect of its idiomatic expression*. Idiomatic expression is therefore not simply an indispensable medium but devoid of merit in its own right and to which recourse is made for lack of better. It is

---

[159]E. Cassirer, *Language and Myth: A Contribution to the Problem of the Names of Gods*, p. 164 (ECW 16, pp. 261–262): "The correlations in being come in accordance with activity, not according to the "objective" similarity of things, but according to the way in which the contents are grasped through the medium of activity and classified together into a determined interconnection of purpose."

[160]E. Cassirer, *The Philosophy of Symbolic Forms*, vol. 1, p. 285 (ECW 11, p. 257): "What primarily distinguished linguistic concept formation from strictly logical concept formation is that it never rests solely on the static representation and comparison of contents but that in it the sheer form of reflection is always infused with specific *dynamic* factors; that its essential impulsions are not taken solely from the world of being but are always drawn at the same time from the world of action."

[161]Cf. supra, § 151.

that by which an activity becomes an object of attention for those who pursue it—
and, in the particular case of linguistic activity, for those who speak a particular
language.[162] Idiomatic expression thus appears as that which attests that language
is the object of a collective pursuit and also as that which is the plane of expression
which must be recognized in its own right for a community of speakers to achieve
self-objectification. We may then better understand what appeared to be the character,
at first glance irrational, of idiomatic expression in the activity of language: because
its relation to the object requires a previous stage of *objectification of the semiotic
activity itself* constituting language into a specific plane of expression, a language can
recruit almost any disparate features to make them into the organized medium of its
own objectification. There is indeed no intrinsically derived and independent reason
for the lexicon of such or such language to favor such or such way or organizing its
own internal coherence, because the random character of its elements does not put
it into question: the idiomatic expression of a language, far from being an obstacle
to its coherence, *manifests on the contrary a process of objectification of its own
form.* It is therefore less a matter of objectivity than a matter of a permanent process
of *objectification* which is at stake with linguistic activity, this process not targeting
what pertains to the concept, but rather the social validation of forms which language
employs to idiomatically signify the invariance and independence identified as the
exteriority of a world.

Language, as a set of forms, thus becomes the object of collective attention only
by instituting itself through an autonomous plane of expression constantly reworked
by speakers in search of a guarantee for the value attributed to the usages they employ.
Speaking then consists in pursuing the work of evaluating forms and of making this
work into the very object of sociality.

3.3   Objectification in Science

Science, in its constant search for determination and for monosemic signification, is
the paradigmatic example of the operator of objectification, as we have abundantly
had the opportunity to analyze in the previous chapters. There is therefore no need,
at this point of our argument, to return to the particular modality of the construction
of its sense. There remains, however, a question to be posed concerning the nature
of the objectivity it deploys and the role played therein by nature when conceived as
independent from the processes of construction which help manifest it. Indeed, as
with the previously cited symbolic forms, the issue of the construction of objectivity
poses at the same time the question of what is not constructed but received, in short,
what we generally conceive as pertaining to "nature".

---

[162]E. Cassirer, *The Philosophy of Symbolic Forms*, vol. 1, p. 277 (ECW 11, p. 248): "This phonetic
symbolism is indicative of that fundamental spiritual process which becomes more and more evident
as language develops. The I grasps itself through its counterpart in verbal action […]."

In the case of science of which the modality of sense construction consists in attempting to overcome its own expressive traces and to extend in accordance with the mode of signification, what serves as foundation appears at first glance to be the strict division between the domain of subjectivity and that of objectivity, and therefore what finds itself cast aside is the relation to sense, only conceived on the mode of the ill-established or of the still undetermined. However, sense is not abolished by science. Cassirer already notes in *Substance and Function* that the conception of nature as a given, independent from knowledge, *has changed over the course of time*: the ancient Platonic "idea" has been replaced by "nature" as conceived by modernity. This has a consequence: on the one hand, if it is the interpretation of expression seen as what must always be sought to be overcome which makes possible the scientific notion of nature, and, on the other hand, if the concept of "nature" is itself likely to evolve, then what makes it accessible is indeed tied with its "constitution", deployed throughout history. Nature is therefore paradoxical in that the independence which science attributes to it *remains problematic*. It is then possible to conceive of the relation of science to nature not only as the ever-reiterable attempt to surpass expression but rather as that which seeks to reintegrate expression into the orbit of rationality on a mode which is not that of language nor of myth. For that which distinguishes science from other symbolic forms, myth and language in particular, is its unique capacity to seek to *consciously* reshape its own relation to sense, thereby getting closer to the mode of productive knowledge, which is that of philosophy. One evident example of this phenomenon, for Cassirer, consists in recognizing in quantum physics a novel attempt to reconstruct the relation between signification and meaning on a whole different basis than that which had until then governed the constitution of modern physics since Galileo. By breaking in particular with the idea of absolute determinism in the way which Emil du Bois-Reymond had exposed the terms,[163] quantum physics attempts to make functional not only the concepts of physics such as those of space-time, but the cognitive act of the physicist as such. It is then a matter of taking into consideration the very conditions of accessibility to what constitutes nature by overcoming the still too idiomatic opposition between subjectivity and objectivity.

The dynamic specific to science, from this perspective, is akin to the philosophy of symbolic forms as Cassirer conceives of it: it is indeed that which seeks to reintegrate the symbolic forms of myth and of language into the orbit of rationality and, from this standpoint, affords science the opportunity to modify its own relation to that

---

[163] E. Cassirer, *Determinism and Indeterminism in Modern Physics; Historical and Systematic Studies of the Problem of Causality*, p. 4 (ECW 19, p. 10): "For Laplace, the idea of this formula was hardly more than an ingenious metaphor by means of which he sought to make clear the difference between the concepts of probability and certainty. The idea that this metaphor should be endowed with a wider meaning and validity, that is should be made the expression of a general epistemological principle, was, to my mind, quite foreign to him. This transition occurred in a much later period, and its date can be established quite definitely. In his speech "Über die Grenzen des Naturerkennens" (1872) [On the limits of the Knowledge of Nature] Emil du Bois-Reymond lifted the Laplacean formula out of its long oblivion and placed it at the focal point of epistemological and scientific discussion."

which it would have otherwise tended to repress, the obscurity of meaning which is however also its source.

3.4 Objectification in Technology

Technology shares with the natural sciences common goals, that is, the tendency to constitute a world governed according to the principles of physical theory, at the level of activity within the sensible world.[164] It is therefore essentially what technology and theoretical science share in common as opposed to the symbolic forms of myth and of language which seems, in Cassirer's view, noteworthy.[165] The question of the independence of nature as it arises when it is question of the operator of objectification, however, takes a specific turn when it is question of technology because it is in respect to the notion of *activity* that the stakes of the debate now lie.

It must first be noted that technology shares the notion of activity with art although these two symbolic forms appear to go in different directions: art seeks above all expression at the level of the sensible whereas technology consists in the "pure form" of the activity. Cassirer notes concerning this difference that the critique which is generally addressed to technology in opposing it to art consists in saying that it essentially amounts to an outwards-directed impulse, whereas art seems to deepen the human subject's activity in all of its dimensions.[166] But for Cassirer, the difference between inside and outside is not itself founded as such and always requires to be reworked in order to make sense.[167]

---

[164] E. Cassirer, "Form and Technology", p. 39 (ECW 17, p. 169): " […] – witness the problem of the flight, which could only finally be solved once technological thinking freed itself from the model of bird flight and abandoned the principles of the moving wing."

[165] E. Cassirer, "Form and Technology", p. 32 (ECW 17, p. 160): "The mythical-magical world still knows nothing about a sense of causality that both constructs and renders possible the sphere of objects, making them accessible to thought. […]. With the creation of the tool and by means of its regular use, the limits of this type of representation were first breached. Here we encounter the 'twilight of the gods' of the magical-mythical world."

[166] E. Cassirer, "Form and Technology", p. 36 (ECW 17, p. 166): "The ludic drive upon which Schiller grounds the region of beauty does not simply add to the mere natural drives such that it would be a broadening of their range, but rather this drive transforms their specific content, first opening up and conquering the proper sphere of 'humanity'. […]. The domain of technological efficacy seems, however, to be denied any such acknowledgment. For, this efficacy appears to be completely subjected by the mastery of those drives which Schiller characterizes as the sentient impulse or as the material drive. The urge towards the outside – that typically 'centrifugal' impulse – manifests itself in it. It brings one piece of the world after another under the dominion of the human will; this spread, this expansion of the periphery of being, thereby leads further and further away from the centre of the 'person' and personal existence."

[167] E. Cassirer, "Form and Technology", p. 37 (ECW 17, pp. 166–167): "Here, Goethe's claim that nature has neither core nor shell rightly applies to the totality of mental activities and energies. Here there is no separation, no absolute barrier between the 'outer' and the 'inner'. Each new gestalt of the world opened up by these energies is likewise always a new opening out of inner existence; it does not obscure this existence, but makes it visible from a new perspective. […]. If we move from

The critique of this rigid point of view concerning the difference between inside and outside actually enlists a whole conception of the subject: far from keeping with the Kantian tradition which keeps with the sole postulation of a transcendental subject, Cassirer acknowledges the movement of recognition and of appropriation of what has been technologically produced as being the *locus of the subject*.[168] We then understand why technology also affords Cassirer the opportunity to present it not as a simply abstract activity, as 'pure a form' as it may be, but as accomplished by subjects[169] conceived to be the mediators of sense. Symbolic forms present themselves as *public* activities which make possible social life defined as the production and indefinite evaluation of forms.

---

this determination, then it would appear at first that knowledge of the I is tied in a very particular sense to the form of technological doing. The border that separates purely organic efficacy from this technological doing is likewise a shard and clear demarcating line with the development of the I-consciousness and singular 'self-knowledge'."

[168] E. Cassirer, "Form and Technology", p. 37 (ECW 17, p. 167): "From the purely physical side, this shows itself in the fact that a determined and clear consciousness of his own body, both a consciousness of his bodily gestalt and his physical functions, first grows in the human being after he turns both of these towards the outside and, so to speak, regains both from the reflection of the outer world."

[169] E. Cassirer, "Form and Technology", p. 43 (ECW 17, p. 174): "In a mind like that of Leonardo da Vinci, the intertwining of these two basic orientations appears with a classic simplicity and depth. What separates Leonardo from mere bookish learning, from the spirit of the 'letterati', as he himself called it, is the fact that the 'theory' and 'praxis', 'praxis' and 'poiesis', penetrate one another in his person in a completely different measure as never before."

Conclusion

To conclude this undertaking, I would first like to insist on a few key points concerning the evolution of Cassirer's thought, an evolution which, as the title of this book indicates, progressively led him from a Kantian-inspired transcendental perspective to a fully semiotic point of view, thereby introducing a whole new manner of conceiving sense and signification.

The central point which profoundly modifies Cassirer's perspective with respect to that of the Kantian tradition concerns the reversal which Cassirer operates upon the latter: whereas Kant, in *The Critique of Judgment*, parsimoniously addressed purely semiotic issues as if they only indirectly had to do with the core of rational discourse, Cassirer on the contrary came to place semiotics at the center of his inquiry regarding the theoretical and practical conditions of knowledge and activities. This reversal was far from being obvious because it required taking elements of knowledge into consideration—knowledge pertaining essentially, at first, to language and myth—which, being replaced at the core of the rational apparatus, still posed a philosophical problem, even if the accumulation of positive data had already, during the nineteenth century, profoundly modified the idea which could be made of the major cultural role played by such semiotic phenomena. In an even more general manner still, by attempting to justify from a philosophical point of view the transformations in knowledge which had developed during the nineteenth and twentieth centuries—historical linguistics and comparative mythology, but also geometry, relativity, and quantum mechanics—transformations of which Kant could obviously not be aware, Cassirer managed to bring about an evolution in the very idea of rationality and to thenceforth distinguish it from the notion of science: the natural sciences lost the commanding role which was theirs until then, all disciplines confounded, all the while continuing to occupy a central position in the particular form of rationality they implemented.

This represents a major change in the very idea of rational inquiry which requires abandoning two key notions of Kantian epistemology: schematism on the one hand and the difference between determinant judgment and reflective judgment on the

© Springer Nature Switzerland AG 2020   183
J. Lassègue, *Cassirer's Transformation: From a Transcendental
to a Semiotic Philosophy of Forms*, Studies in Applied Philosophy, Epistemology
and Rational Ethics 55, https://doi.org/10.1007/978-3-030-42905-8

other. Schematism, in its Kantian version, does not allow to elaborate a fully functional theory of concepts and hence remains prisoner of a theory of the true reflection of the object without the functional co-belongingness of the theory and the object being fully thinkable. With Cassirer, the role which was first attributed to schematism is ensured by the *intrinsically organizing value of the relation between sense and signification in the concepts forming a theory* and it is precisely this point which introduces a new intelligibility of what he calls "symbolic". As for the distinction between determinant judgment and reflective judgment, it bears the inconvenience of instituting a radical separation between the natural sciences and the humanities, the former being supposed to make possible the constitution of objectivity whereas the latter, devoid of a constitutive role, would only manifest the intrinsic power of the subject when left without the aid of schematism. Now, the research conducted during the nineteenth and twentieth centuries have, in Cassirer's view, profusely shown that, on the one hand, the humanities are no less "determinant" than the natural sciences if they are seen as developing certain modalities of the relation between signification and sense and if, on the other hand, the natural sciences make use of concepts and methods which would have tended to be spontaneously declared "reflective", in particular with respect to the use they make of the notions of totality and of form. It is therefore not there where their difference lies, but in the *dynamic specific to their mode of objectification*, that is, in the differentiated relation they have with the expressive basis they all share: whether they conceive of the relation to expression as requiring to be overcome or as having intrinsic generative value, according to the conscious and unconscious role they attribute to signs, or whether they conceive of the relation to meaning in terms of persistence or of divorce, knowledge progressively divides itself into branches which finish by mutually excluding one another, but which can also intersect once more following trajectories that cannot be anticipated but which can be described using a cluster of categorical operators which are themselves likely to evolve. These remarks thus enable to outline the project of a philosophy of symbolic forms which continues today to lend itself to many borrowings and reformulations.

By themselves, these could provide matter for a whole book about the topicality and the future of Cassirer's philosophy, but three among them deserve to be emphasized because they intersect difficulties which were my own during the writing of this book. Exposing them can not only make the reader's interrogations appear more legitimate if they are shared during the reading but may also incite the reader to pursue in his or her own manner the reflection which remains open.

The first, which confronts anyone who would propose to clarify or to emphasize the relevance of a philosopher's work, consists in keeping as close as possible with a corpus of texts all the while attempting to reveal even that which went unthought. Cassirer's case is no exception: both effacement and intervention is required from the part of one who wishes to clarify Cassirer's thinking, whether this thinking has already fully matured or if it is still in search of itself. Seeing a certain tour de force in the interpretation which uncovers the activity of the interpreter instead of keeping it in the shadows is not something negative and can encourage the reader to take position in the debate and to make his or her own attempt at the art of interpretation.

This represents however only of a general difficulty which is not exclusive to the interpretation of the work of Cassirer. Two more specific difficulties arise when attempting to evaluate Cassirer's work as a whole.

I have insisted on several occasions throughout the preceding pages upon a capital problem which appears in the work of Cassirer, that is, the ambivalent status surrounding the notion of sign: at first conceived along the transparent and univocal terms favored by a scientific perspective of which the purview is universal from the onset, its status evolves and becomes more differentiated, while the preconceived idea of a homogeneity in the deployment of symbolic forms is put into question, that is, while the idea of a universal schematism tends to efface itself. Then the idea of true individuation specific to each symbolic form arises, deploying specific values and thereby requiring an intrinsic evaluation grid which is adapted to each. The problem of the ambivalence of the status of the sign is not however further thematized in a direct manner by Cassirer, even if the whole movement of his work gravitates around this question. How is this status to be conceived, that is, how to conceive of the intrinsic variability deployed by signs in order to remain themselves? The answer I have attempted to formulate while keeping with what I understood of Cassirer's thought seems to revolve around the processes of individuation of significations in their relation to the instituted ground of sense.

This first answer has inflected the course of my reading of Cassirer's work with respect to a second point, that of the place to confer to language among the whole set of symbolic forms. Such a place seems ambiguous to me due to the ambivalence of the status of the sign of which it has just been question. By keeping with the perspective put forth in *The Philosophy of Symbolic Forms*, and particularly with its first volume, I first conceived of language as a sort of *princeps* symbolic form, that to which it was always necessary to return in order to account both for the expressive background of sense in general and for the possibility of making its pregnancy vary in accordance with the various forms of signification. This interpretation was finally confronted with two difficulties. On the one hand, it did not enable to account for the relation which language could have with phenomena that are obviously symbolic but not directly linguistic, such as technology. On the other hand, it tended to skew the issue of the role of language in a "pre-critical" direction by relating this role to the metaphysical problem of foundations. However, that language must necessarily be involved when posing the foundational question regarding symbolic forms, does not mean that the answer to this question must consecrate language as a foundation, but only as that which enables to conceive the foundation. It is this way of looking at the both central and indirect role of language which incited me to adopt a broader point of view rather than an exclusively linguistic one regarding the nature of the symbolic processes such as they are described by Cassirer in his ulterior works. Regarding these two capital questions, it is up to the reader to evaluate my answers and, if required, to question them.

A last point deserves, in my view, to be emphasized regarding the place occupied by Cassirer's work in today's philosophical reflection regarding the notion of culture. This point is of a *political* nature, taken in the broadest sense. The destiny of the reception of Cassirer's work has been in part shadowed by the dramatic circumstances in which it had been produced: being a German, republican, and Jewish philosopher, his destiny was sealed by Nazism and the exile he underwent from 1933 until his death in 1945—an exile to Great-Britain, Sweden, and finally the United States—certainly had repercussions regarding the way in which he worked and the way in which his works have been read. Re-reading Cassirer today *consists also in reviving the ties with the German classical tradition beyond Nazism*, and it is in this respect that this re-reading has political consequences, consequences which do not solely concern German culture: the problem of what lies beyond Kantianism has indeed elicited several international contributions—it is, in particular, at the origin of the works of the French school of sociology founded by Durkheim—even if it concerns first and foremost German culture and its future. In what concerns our current situation, we know the preponderant role that analytical philosophy of a firstly Anglo-American logicist inspiration has progressively carved out for itself in all academic environments and, from this perspective, one might consider the international rebirth of Cassirerean studies to be stillborn due to their apparent remoteness from the issues and questions raised by the analytical tradition. However, since the latter renounced its frontal opposition to all questions of a semiotic nature and that it embarked upon a search for a philosophy of culture, Cassirer's work may be viewed in new light and contribute in reviving the ties with a response to the problem of culture in which the notion of limit is conceived as the primary modality of the relation to universality. Philosophically speaking, the reciprocal limitation of forms of which the tendency is to integrally saturate the field of sense then makes of self-limitation the path to universality. Cassirer's philosophy can thus enable to renew the idea of an intellectual cosmopolitanism the boundaries of which end where the problem of cultural variability as the pathway to universality ceases to constitute the very theme of social life. There is therefore a *topicality* in the philosophy of Cassirer which should not only be recognized, but also promoted. This book, I hope, will have contributed somewhat to this.

Bibliography

Ernst Cassirer Reference Edition

(i) *Gesammelte Werke*, Hamburger Ausgabe, Felix Meiner Verlag, 1998–2009, appearing in footnotes as '**ECW**' followed by the page number.

(ii) *Ernst Cassirer Nachgelassene Manuskripte und Texte*, Felix Meiner Verlag, 1995–2017, appearing in footnotes as '**ECN**' followed by the page number.

Cited Texts by Cassirer

[1906 & 1910–1911] *Das Erkenntnisproblem in der Philosophie und Wissenschaft der neueren Zeit*, Band 1, (**ECW** 2).

[1907] „Kant und die moderne Mathematik", *Kant-Studien*, 12, pp. 1–40, (**ECW** 9, pp. 37–82).

[1907] *Das Erkenntnisproblem in der Philosophie und Wissenschaft der neueren Zeit*, Band 2, (**ECW** 3).

[1910] *Substance and Function*, The Open Court Publishing Company, Chicago, 1923, (**ECW** 6).

[1912] "Hermann Cohen und die Erneuerung der Kantischen Philosophie", *Kant-Studien* 17 (1912) pp. 252–273 (**ECW** 9, pp. 119–138).

[1916] *Freiheit und Form; Studien zur Deutschen Geistesgeschichte*, (**ECW** 7).

[1921] *Einstein's Theory of Relativity*, The Open Court Publishing Company, Chicago, 1923, (**ECW** 10).

[1921] „Goethe und die mathematische Physik. Eine erkenntnistheoretische Betrachtung", (**ECW** 9, pp. 268–315).

[1923] *The Philosophy of Symbolic Forms*, vol. 1, trans. by Ralph Manheim, New Haven & London, Yale University Press, 1955 (**ECW** 11).

[1923] "The Concept of Symbolic Form in the Construction of the Human Sciences", Vorträge der Bibliothek Warburg, 1 (Leipzig: B. G. Teubner, 1922), 11–39, trans. in *E. Cassirer, The Warburg Years (1919–1933)*, pp. 72–100 (**ECW** 16, pp. 75–104).

[1924] "Eidos and Eidolon: The Problem of Beauty and Art in the Dialogues of Plato" Vorträge der Bibliothek War-burg, 2 (Leipzig: B. G. Teubner, 1924), 1–27 trans. in E. Cassirer, *The Warburg Years (1919–1933)*, pp. 214–243 (**ECW 16**, pp. 135–163).

[1925] *The Philosophy of Symbolic Forms*, volume 2, trans. by Ralph Manheim, New Haven & London, Yale University Press, 1955 (**ECW** 12).

© Springer Nature Switzerland AG 2020

J. Lassègue, *Cassirer's Transformation: From a Transcendental to a Semiotic Philosophy of Forms*, Studies in Applied Philosophy, Epistemology and Rational Ethics 55, https://doi.org/10.1007/978-3-030-42905-8

[1925] *Language and Myth: A Contribution to the Problem of the Names of Gods* in *E. Cassirer, The Warburg Years (1919–1933)*, pp. 130–213 (**ECW** 16, 227–311).

[1927] *The Individual and the Cosmos in Renaissance Philosophy*, Philadelphia, University of Pennsylvania Press, 1972 (**ECW 14**, pp. 1–220).

[1927] "The Problem of the Symbol and Its Place in the System of Philosophy", *Zeitschrift für Ästhetik und allgemeine Kunstwissenschaft* 21; Stuttgart: Enke, 1927, pp. 191–208 trans. in *E. Cassirer, The Warburg Years (1919–1933)*, pp. 254–271 (**ECW** 17, pp. 227–311).

[1929] The Davos Debate in *European Existentialism*, ed. Nino Langiulli, New Brunswick, N.J.: Transaction, 1997, pp. 202–203 (**ECN 17**).

[1929] *The Philosophy of Symbolic Forms*, vol. 3, trans. by Ralph Manheim, New Haven & London: Yale University Press, 1955 (**ECW 13**).

[1930], "Form and Technology", in *Kunst und Technik*, ed. Leo Kestenberg, Berlin: Volksverband der Bücherfreude-Verband, pp. 15–61, trans. by W. McClelland Dunlavey & J.M. Krois, in *Ernst Cassirer on Form and Technology; Contemporary Readings*, Hoel A.S. & Folkvord I. (eds.), Palgrave, Macmillan, 2012, pp. 15–53 (**ECW 17**, pp. 139–183).

[1931] "Mythic, Aesthetic and Theoretical Space", in *Vierter Congress für Ästhetik und allgemeine Kunstwissenschaft*, ed. H. Noack (Stuttgart: Enke, 1931), 21–36, trans. in *Ernst Cassirer; The Warburg Years (1919–1933)*, translated by S. G. Lofts with A. Calagno, Yale University Press, New Haven and London, 2013, pp. 317–333 (**ECW 17**, pp. 411–432).

[1932] *The Platonic Renaissance in England*, translated by James P. Pettegrove, Nelson, 1953, (**ECW 14**, pp. 223–380).

[1936] *Determinism and Indeterminism in Modern Physics: Historical and Systematic Studies of the Problem of Causality*, (**ECW 19**).

[1939] *Axel Hägerström. Eine Studie zur Schwedischen Philosophie der Gegenwart*, (**ECW 21**, pp. 3–116).

[1940] *The Problem of Knowledge; Philosophy, Science, and History since Hegel*, transl. by William H. Woglom & Charles W. Hendel, New Haven, Yale University Press, 1950 (**ECW 5**).

[1940] „Geist und Leben" (**ECN 1**, pp. 3–32; pp. 207–228).

[1940] „Das Symbolproblem als Grundproblem der philosophischen Anthropologie" in *The Philosophy of Symbolic Forms*, vol. 4, trad. J. Krois & D. Verene, Yale University Press, 1996, (**ECN 1**, pp. 32–109).

[1940] „Über Basisphänomene" in *The Philosophy of Symbolic Forms*, vol. 4, trad. J. Krois & D. Verene, Yale University Press, 1996 (**ECN 1**, pp. 113–195).

[1942] "The Influence of Language upon the Development of Scientific Thought", *The Journal of Philosophy*, vol. 39, n 12, pp. 309–327, (**ECW 24**, pp. 115–134).

[1942]—"The Object of the Science of Culture", in Ernst Cassirer, *The Logic of the Cultural Sciences*, Yale University Press, 1960, pp. 29–30 (**ECW 24**, pp. 357–390).

[1942]—"The problem of Form and the Problem of Cause" in *The Logic of the Humanities*, Yale University Press, 1960, pp. 179–180 (**ECW 24**, pp. 446–461).

[1944] "The Concept of Group and the Theory of Perception", *Philosophy and Phenomenological Research*, vol. 5, no 1, pp. 1–36 (**ECW 24**, pp. 209–250).

[1944]—*An Essay on Man, An Introduction to a Philosophy of Human Culture*, **ECW 23**.

[1945]—"Structuralism in Modern Linguistics", *Word*, 1, August 1945, pp. 97–120 (**ECW 24**, pp. 299–320).

[1945]—"Goethe and the Kantian Philosophy" in *Rousseau, Kant, Goethe, Two Essays*, Princeton 1945, The History of Ideas Series, vol. 1, (**ECW 24**, pp. 542–575).

[1946], —*The Myth of the State*, (**ECW 25**).

Other Texts

Aristotle, *Posterior Analytics*, Clarendon Press, Oxford, 1994.

Brinton D. G., *Religions of Primitive Peoples*, G. P. Putnam's Sons, New York 1897.

Bühler K., *Theory of Language; The representational function of language*, John Benjamins Publishing Company, 2011.

Cohen H., *Ästhetik des reinen Gefühls*, Band 1, B. Cassirer, Berlin, 1912.

Curtius G., *Grundzüge der griechischen Etymologie*, Teubner, Leipzig, 1873.

Darwin C., *The Expression of the Emotions in Man and Animals*, Oxford University Press, Oxford, 2009.

Duhem P., *The Aim and Structure of Physical Theory*, New York, Atheneum, 1962; reprint Princeton, Princeton University Press, 1991.

du Bois-Reymond E., "Über die Grenzen des Naturerkennens : Ein Vortrag in der 2. offentlichen Sitzung der 45. Versammlung deutscher Naturforscher und Arzte zu Leipzig am 14. August 1872", ReInk Book, New Deli, 2016.

Euclid, *The Thirteen Books of Euclid's Elements*, transl. T. L. Heath, University Press, Cambridge, 1908.

Ferrari M., *Ernst Cassirer; dalla scuola di Marburgo alla filosofia della cultura*, Leo S. Olschki Editore, Firenze, 1996.

Frege G., *Philosophical and Mathematical Correspondence*, ed. By G. Gabriel, H. Hermes, F. Kambartel, C. Thiel, A. Veraart, trans. By H. Kaal, The University of Chicago Press, 1980.

Freud S., *Totem and Taboo*, Freud Press, 2013.

Galileo, —*Dialogue Concerning the Two Chief World Systems*, Moder Library, 2001.

Galileo, —*Il Saggiatore*, Milan, Feltrinelli, 2008.

Gawronsky D., "Ernst Cassirer: His Life and Work", in *The Library of Living Philosophers* vol. 6, P. A. Schilpp ed., Northwestern University, 1949.

Habermas J., „Die befriende Kraft der symbolischen Formgebung", in D. Frede and R. Schmücket (des), *Ernst Cassierers Werk und Wirkung*, Darmstadt, Wiss. Buchges., 1997.

Hegel G.W.F., "Vorlesungen über die Philosophie der Geschichte", *Sämmtliche Werke* (Leipzig, 1949), 9, 98.

Hilbert D., *Foundations of Geometry*, Merchant Books, 2007.

Hilbert D., „Neubegründung der Mathematik", *Gesammelte Abhandlungen*, Band III, Springer Verlag, 1970, pp. 156–177.

Iribarren L., "Langage, mythe et philologie dans la *Philosophie des formes symboliques* d'Ernst Cassirer" *Revue germanique internationale*, 15/02, 2012, pp. 95–114.

Kant E., —"Von dem ersten Grunde des Unterschieds der Gegenden im Raume" [1768] in E. Cassirer (ed.), *Immanuel Kants Werke*, Band 2.

Kant E., —*The False Subtlety of the Four Syllogistic Figures Proved*, Barnes & Noble, 2005.

Kant E., —*Prolegomena to Any Future Metaphysics That Will Be Able to Present Itself as a Science*, Oxford University Press, Oxford, 2003.

Kant E., —*Critique of Pure Reason*, Hackett Publishing, 1996.

Kant E., —*Über die Deutlichkeit der Grundsätze der natürlichen Theologie und der Moral* [http://gutenberg.spiegel.de/buch/-6399/1].

Kant E., —*Critique of Judgement*, MacMilland & Co., London, 1914.

Klein F., —"A comparative Review of recent Researches in Geometry" (1872), English translation by Dr. M. W. Haskell and transcribed by N. C. Rughoonauth, *Bull. New York Math. Soc.* 2, (1892–1893) [http://arxiv.org/abs/0807.3161].

Klein F., —"Erlanger Programm; Vergleichende Betrachtungen über neuere geometrische Forschungen", *Mathematische Annalen*, 43, 1893 [http://math.ucr.edu/home/baez/erlangen/].

Koffka K., *Die Grundlagen der psychischen Entwicklung*, (1921) [https://archive.org/details/diegrundlagende00koffgoog].

Köhler, W. (1922), „Zur Psychologie der Schimpansen", in *Psychologische Forschung. Zeitschrift für Psychologie und ihre Grenzwissenschaften* 1, 2–46.

Köhler W., „Komplextheorie und Gestalttheorie", *Psychologische Forschung. Zeitschrift für Psychologie und ihre Grenzwissenschaften* 6, 358–416.

Koyré A., *From the Closed World to the Infinite Universe*, Harper, 1958.

Krois J. M., "The priority of "symbolism" over language in Cassirer's philosophy", *Synthese*, 2008, 179, pp. 9–20.

Krois J. M., "Ernst Cassirer's philosophy of biology", *Sign Systems Studies* 32.1/2, 2004: 277–295.

Krois J. M., "Symbolisme et phénomène de base (*Basisphänomene*)", *Revue Germanique Internationale*, 15/2012, pp. 161–174.

Kreisel G., "Hilbert's Programme", *Dialectica*, Volume 12, Issue 3–4, Dec. 1958, pp. 346–372.

Legendre A. M., *Éléments de géométrie*, Firmin Didot Frères, Paris, 1862.

Lévy-Bruhl L., *Les fonctions mentales des sociétés inférieures*, Presses Universitaires de France, Paris, 1910.

Leibniz G. W., *Monadology*, Hackett Publishing Company, 1991.

Lotze H., *Metaphysics*, Oxford, Clarendon Press, 1887.

Pierobon F., *Kant et les mathématiques; la conception kantienne des mathématiques*, Vrin, Paris, 2003.

Poincaré H., *Science and Hypothesis*, New York, Dover Publications, 1952.

Recki R., "Cassirer and the problem of language" in *Cultural Studies and the Symbolic* ed. By Paul Bishop and R. H. Stephenson, Northern Universities Press, 2003.

Riemann B., "Über die Hypothesen, welche der Geometrie zu Grunde liegen", *Abhandlungen der Königlichen Gesellschaft der Wissenschaften zu Göttingen*, 1867, Band 13, pp. 133–150; English transl. "On the Hypotheses which lie at the Bases of Geometry", *Nature*, vol. VIII.

Rosenthal V. & Visetti Y.-M., *Koehler*, Les Belles Lettres, Paris, 2003.

Stjernfelt F., "Simple Animals and Complex Biology. The double von Uexküll inspiration in Cassirer's philosophy", in *Synthese*, Vol. 179, No. 1, pp. 169–186, 2009.

Steinthal, *Die Mande-Neger-Sprachen*, Ferd. Dümmler's Verlag Buchhandlung, Berlin, 1867 [https://archive.org/details/diemandenegersp01steigoog].

Verene D. P., *Symbol, Myth and Culture; Essays and Lectures of Ernst Cassirer 1935–1945,* Yale University Press, 1979.

Von Humboldt W. „Einleiting zum Kawi-Werk", *Werke*, 7, No 1.

Weyl H., —*Gruppentheorie und Quantenmechanik*, Hirlzel, Lepizig, 1931; trans. *The Theory of Groups and Quantum Mechanics*, E. P. Dutton, New York, 1932.

Weyl H., —"Philosophie der Mathematik und Naturwissenschaft", *Handbuch der Philosophie*, Munich & Berlin, 1927; trans. *Philosophy of Mathematics and Natural Science*, Princeton, Princeton University Press, 2009.

Wundt W., „der mathematische Raumbegriff", *Logik*, I, p. 496, 2. Aufl. Stuttgart, Ferdinand Enke, 1893.

Wussing H., *The Genesis of the Abstract Group Concept : a Contribution to the History of the Origin of Abstract Group Theory,* Cambridge, MIT Press, 1984.

Printed in the United States
by Baker & Taylor Publisher Services